Lecture Notes in Computer Science 15410

Founding Editors

Gerhard Goos
Juris Hartmanis

Editorial Board Members

Elisa Bertino, USA
Wen Gao, China
Bernhard Steffen, Germany
Moti Yung, USA

FoLLI Publications on Logic, Language and Information

Subline of Lecture Notes in Computer Science

Subline Editors-in-Chief

Valentin Goranko, *Stockholm University, Sweden*
Michael Moortgat, *Utrecht University, The Netherlands*

Subline Area Editors

Nick Bezhanishvili, *University of Amsterdam, The Netherlands*
Anuj Dawar, *University of Cambridge, UK*
Philippe de Groote, *Inria Nancy, France*
Gerhard Jäger, *University of Tübingen, Germany*
Fenrong Liu, *Tsinghua University, Beijing, China*
Eric Pacuit, *University of Maryland, USA*
Ruy de Queiroz, *Universidade Federal de Pernambuco, Brazil*
Ram Ramanujam, *Institute of Mathematical Sciences, Chennai, India*

Jialiang Yan · Mingming Liu · Dag Westerståhl ·
Xiaolu Yang
Editors

The Connectives in Logic and Language

4th Tsinghua Interdisciplinary Workshop on Logic, Language, and Meaning, TLLM 2024 Beijing, China, March 29–31, 2024 Revised Selected Papers

Springer

Editors
Jialiang Yan
Tsinghua University
Beijing, China

Dag Westerståhl
Stockholm University
Stockholm, Sweden

Tsinghua University
Beijing, China

Mingming Liu
Tsinghua University
Beijing, China

Xiaolu Yang
Tsinghua University
Beijing, China

ISSN 0302-9743 ISSN 1611-3349 (electronic)
Lecture Notes in Computer Science
ISBN 978-3-031-86053-9 ISBN 978-3-031-86054-6 (eBook)
https://doi.org/10.1007/978-3-031-86054-6

This Springer imprint is published by the registered company Springer Nature Switzerland AG
The registered company address is: Gewerbestrasse 11, 6330 Cham, Switzerland

Preface

The propositional connectives – and, or, not, if–then, etc. – are fundamental building blocks in formal as well as natural languages. In the Western tradition, they were first studied as such by the Stoics, and Propositional Logic is the fundament of practically all current systems of logic; every beginning logic course starts with it. Still, the proof theory and semantics of systems of propositional logic are far from trivial, and have been studied intensely by logicians in the last one and a half centuries, not least in recent decades. It is actually a vast area of research, as witnessed by Lloyd Humberstone's 1500 page tome *The Connectives* (2011), which overviews much of that research. Perhaps the most familiar recent work in this area concerns conditionals in formal and natural languages. The 4th Tsinghua Interdisciplinary Workshop on Logic, Language, and Meaning (TLLM 2024) also focused on the apparently simpler connectives expressing (various versions of) conjunction, disjunction, and negation.

Researchers working from a cross-linguistic perspective also study how the connectives are encoded in different languages, and ask whether classical logic is capable of capturing the variations and universals exhibited. Even in well-studied languages such as English, there are intricate phenomena that remain challenging for classical logic, including free choice disjunction, non-Boolean conjunction, and metalinguistic negation, to name just a few. There is also growing interest in the acquisition and processing of natural language connectives. In the context of the hotly discussed Large Language Models (LLMs), understanding connectives presents novel challenges that deserve in-depth exploration.

The TLLM workshops bring together logicians and linguists around a specific theme of common interest. Previous themes have been, respectively, Exceptives/Monotonicity/Dynamics in Logic and Language. The 2024 workshop was the first to be held with personal attendance since the COVID pandemic. It took place March 29–31, on site at Tsinghua University, and was attended by around 50 scholars and students.

Forty-nine abstracts were submitted to the workshop, 18 of which were presented as contributed talks, and 7 at poster sessions. In addition, there were four invited talks and two tutorials. One of the invited papers is published in this volume, together with seven of the contributed papers, after two carefu single-blind reviews per paper. We express our sincere gratitude to all the colleagues who helped to review the submissions.

This was the fourth edition of the workshop series Interdisciplinary Workshops on Logic, Language, and Meaning, held at Tsinghua University since its successful debut in April 2019. The plan is to continue the event and keep exploring further aspects of the interface between logic and language.

December 2024

Jialiang Yan
Mingming Liu
Dag Westerståhl
Xiaolu Yang

Organization

Tutorials

Connectives in Logic, by Wesley Holliday, UC Berkeley, USA
Connectives in Language, by Christoph Harbsmeier, University of Oslo, Norway

Invited Speakers

Wesley Holliday	UC Berkeley, USA
Christoph Harbsmeier	University of Oslo, Norway
Jacopo Romoli	Heinrich Heine University Düsseldorf, Germany
Fan Yang	University of Utrecht, The Netherlands

Program Committee

Maria Aloni	University of Amsterdam
Gennaro Chierchia	Harvard University
Xuping Li	Zhejiang University
Jo-wang Lin	Academia Sinica
Fenrong Liu	Tsinghua University
Mingming Liu (Co-chair)	Tsinghua University
Larry Moss	Indiana University Bloomington
Stanley Peters	Stanford University
Martin Stokhof	University of Amsterdam and Tsinghua University
Jakub Szymanik	University of Trento
Johan van Benthem	Stanford University and University of Amsterdam and Tsinghua University
Frank Veltman	University of Amsterdam
Yingying Wang	Hunan University
Dag Westerståhl (Co-chair)	Stockholm University and Tsinghua University
Yicheng Wu	Zhejiang University
Xiaolu Yang (Co-chair)	Tsinghua University
Linmin Zhang	NYU Shanghai

Organizing Committee

Jialiang Yan (Chair)	Tsinghua University, China
Wei Wang	Tsinghua University, China
Rong He	Tsinghua University, China
Qingyu He	Tsinghua University, China
Penghao Du	Tsinghua University, China

Sponsors

Joint Research Center for Logic, Tsinghua University, China
Department of Philosophy, Tsinghua University, China
Department of Foreign Languages and Literatures, Tsinghua University, China

Contents

Vagueness and the Connectives 1
Wesley H. Holliday

A Probabilistic Logic for Causal Counterfactuals 23
Jingzhi Fang

Disjunction, Juxtaposition, and Alternative Questions in Mandarin 51
Rong He

Disjunctions of Universal Modals and Conditionals 69
Dean McHugh

Questions and Connectives 93
Tue Trinh and Itai Bassi

Variation in Conditional Perfection: A Comparative Study
of English/German Versus Mandarin Chinese 106
Alexander Wimmer and Mingya Liu

Disjunction and Parentheses 127
Jialiang Yan, Chen Ju, and Wei Wang

Negation, Disjunction, and Choice Questions: Reflections on Yuen Ren
Chao on Chinese Logical Expressions 149
Byeong-uk Yi

Author Index 189

Vagueness and the Connectives

Wesley H. Holliday(✉)

University of California, Berkeley, USA
wesholliday@berkeley.edu

Abstract. Challenges to classical logic have emerged from several sources. According to recent work [24], the behavior of *epistemic modals* in natural language motivates weakening classical logic to orthologic, a logic originally discovered by Birkhoff and von Neumann [3] in the study of quantum mechanics. In this paper, we consider a different tradition of thinking that the behavior of *vague predicates* in natural language motivates weakening classical logic to intuitionistic logic or even giving up some intuitionistic principles. We focus in particular on Fine's recent approach to vagueness [12]. Our main question is: what is a natural non-classical base logic to which to retreat in light of both the non-classicality emerging from epistemic modals and the non-classicality emerging from vagueness? We first consider whether orthologic itself might be the answer. We then discuss whether accommodating the non-classicality emerging from epistemic modals and vagueness might point in the direction of a weaker system of fundamental logic [21].

Keywords: Vagueness · non-classical logic · orthologic · intuitionistic logic · compatibility logic · fundamental logic

1 Introduction

At the heart of one formulation of the Sorites Paradox is the classical inconsistency of

$$\underbrace{p_1 \wedge \neg p_n}_{\text{extremes}} \wedge \underbrace{\neg(p_1 \wedge \neg p_2) \wedge \neg(p_2 \wedge \neg p_3) \wedge \cdots \wedge \neg(p_{n-1} \wedge \neg p_n)}_{\text{no sharp cutoff}}. \tag{1}$$

For example, for a vague predicate like 'young', we would like to say that a person who has been alive for 1 s is young and that a person who has been alive for 3,000,000,000 s[1] is not young, but there is no exact k such that someone who has been alive for k seconds is young while someone who has been alive for $k+1$ seconds is not young. Yet the combination of these claims as in (1) is inconsistent in classical logic.

[1] Approximately 95 years.

J. Yan et al. (Eds.): TLLM 2024, LNCS 15410, pp. 1–22, 2025.
https://doi.org/10.1007/978-3-031-86054-6_1

In fact, (1) is even inconsistent in *intuitionistic* logic.[2] Those who have thought that the Sorites Paradox motivates moving from classical to intuitionistic logic (e.g., [4,34]) are apparently satisfied that while

$$\textsf{extremes} \vdash_{\textsf{Int}} \neg\textsf{no sharp cutoffs},$$

at least $\neg$no sharp cutoff does not intuitionistically entail that there is some sharp cutoff:

$$\neg\textsf{no sharp cutoff} \nvdash_{\textsf{Int}} \underbrace{(p_1 \wedge \neg p_2) \vee \cdots \vee (p_{n-1} \wedge \neg p_n)}_{\textsf{some sharp cutoff}}.$$

Yet this subtle position does not satisfy those who would like to make the assertions after (1) above.

By contrast, Fine [12] develops an approach to the Sorites that renders (1) *consistent.* In a fragment of his language with just $\wedge$, $\vee$, and $\neg$, Fine's logic is the logic of bounded distributive lattices equipped with a so-called weak pseudocomplementation (Proposition 4 below).[3] Over such algebras, (1) is satisfiable. However, requiring these algebras to also satisfy the law of excluded middle, $a \vee \neg a = 1$, turns them into Boolean algebras and hence yields classical logic, rendering (1) inconsistent. Thus, the rejection of the law of excluded middle is essential to Fine's approach to the Sorites, as it was to Field's [11].

While Fine's logic of vagueness includes the distributivity of $\wedge$ and $\vee$, Holliday and Mandelkern [24] have recently argued that distributivity is not generally valid, using examples involving the epistemic modals 'might' and 'must'. They specifically argue for the adoption of *orthologic*, a logic originally discovered by Birkhoff and von Neumann [3] in the study of quantum mechanics, as the underlying logic for $\wedge$, $\vee$, and $\neg$ in a language with epistemic modals. However, they do not consider whether vagueness might motivate weakening orthologic still further, e.g., by dropping the law excluded middle—which is included in orthologic—when admitting vague predicates into the language. This is the question we wish to address in this paper: what is a natural non-classical base logic to which to retreat in light of both the non-classicality emerging from epistemic modals and the non-classicality emerging from vagueness?

The first candidate we will consider is simply orthologic itself. As we will see, like Fine's logic, orthologic renders (1) consistent and does so with quite intelligible semantic models, but it does so thanks to the denial of distributivity rather than excluded middle; though $(p_1 \vee \neg p_1) \wedge \cdots \wedge (p_n \vee \neg p_n)$ is an orthological theorem applicable to the Sorites Paradox, orthologic does not permit the

[2] This is easy to see using the standard preorder semantics for intuitionistic logic [7,18,26]: $\neg(p_k \wedge \neg p_{k+1})$ is forced at a state x iff for every successor x' of x that forces p_k, there is a successor x'' of x' that forces p_{k+1}, which is a successor of x by transitivity. Thus, if x forces p_1, then applying the previous reasoning to each conjunct of no sharp cutoff, it follows that there is a successor of x that forces p_n, so $\neg p_n$ cannot also be forced at x.

[3] A unary operation $\neg$ on a bounded lattice L is a *weak pseudocomplementation* if for all elements a, b of L, $a \wedge \neg a = 0$, $a \leq \neg\neg a$, and $a \leq b$ implies $\neg b \leq \neg a$. Equivalently, for all elements a, b of L, $a \wedge \neg a = 0$, and $a \leq \neg b$ implies $b \leq \neg a$.

distributive inference from that conjunction of disjunctions to the disjunction of all conjunctions of the form $\pm p_1 \wedge \cdots \wedge \pm p_n$, where $\pm$ is empty or $\neg$, which with $p_1 \wedge \neg p_n$ and distributivity implies some sharp cutoff. This may seem like an attractively parsimonious solution: perhaps weakening classical logic to orthologic is sufficient to handle both epistemic modals and vagueness. However, there are further desiderata for the logic of vagueness beyond making (1) consistent, and perhaps these further desiderata motivate giving up excluded middle after all.

This brings us to another candidate, the system we call *fundamental logic* [21]. Fundamental logic weakens orthologic precisely by dropping the principle of double negation elimination that gives rise to excluded middle, as the phenomena involving vagueness might motivate us to do, and weakens Fine's logic precisely by dropping distributivity, as the phenomena involving epistemic modals might motivate us to do. In [21], fundamental logic is defined in terms of a Fitch-style natural deduction system including only introduction and elimination rules for the connectives $\wedge$, $\vee$, and $\neg$. In particular, this natural deduction system does not include the rule of Reductio Ad Absurdum (if the assumption of $\neg\varphi$ leads to a contradiction, then conclude φ) or the rule that Fitch called Reiteration (which allows pulling certain previously derived formulas into subproofs). Modulo the introduction and elimination rules, Reductio Ad Absurdum is the culprit behind double negation elimination and excluded middle, while Reiteration is the culprit behind distributivity. By including only the introduction and elimination rules for the connectives, fundamental logic avoids those principles and the associated problems involving vague predicates and epistemic modals.

The rest of the paper is organized as follows. In § 2, we define the logics that we will consider as candidate logics for a language with $\wedge$, $\vee$, $\neg$, and vague predicates. In § 3, we briefly review a general relational semantics for orthologic, fundamental logic, intuitionistic logic, and classical logic. In § 4, we review a different semantics that Fine uses for the purposes of his logic. We then turn to *models* of the Sorites series: first a model appropriate for orthologic or Fine's logic (depending on the semantics applied to the model) in § 5 and then a model appropriate for fundamental logic in § 6. These models show that orthologic and fundamental logic, which can handle the non-distributivity arising from epistemic modals, can also handle the Sorites series at least in some minimal sense. In § 7, we discuss whether we should favor an orthological approach to the Sorites or a fundamental one. We conclude in § 8 with some open questions.

2 Logics

We will focus on the logical connectives of negation, conjunction, and disjunction, using the following formal language.

Definition 1. Given a nonempty set Prop of propositional variables, let $\mathcal{L}$ be the propositional language generated by the grammar

$$\varphi ::= p \mid \neg\varphi \mid (\varphi \wedge \varphi) \mid (\varphi \vee \varphi)$$

where $p \in \mathsf{Prop}$.

The logics we will consider for this language are all examples of "intro-elim logics" in the following sense from [21].

Definition 2. An *intro-elim logic* is a binary relation $\vdash \subseteq \mathcal{L} \times \mathcal{L}$ such that for all $\varphi, \psi, \chi \in \mathcal{L}$:

1. $\varphi \vdash \varphi$
2. $\varphi \wedge \psi \vdash \varphi$
3. $\varphi \wedge \psi \vdash \psi$
4. $\varphi \vdash \varphi \vee \psi$
5. $\varphi \vdash \psi \vee \varphi$
6. $\varphi \vdash \neg\neg\varphi$
7. $\varphi \wedge \neg\varphi \vdash \psi$
8. if $\varphi \vdash \psi$ and $\psi \vdash \chi$, then $\varphi \vdash \chi$
9. if $\varphi \vdash \psi$ and $\varphi \vdash \chi$, then $\varphi \vdash \psi \wedge \chi$
10. if $\varphi \vdash \chi$ and $\psi \vdash \chi$, then $\varphi \vee \psi \vdash \chi$
11. if $\varphi \vdash \psi$, then $\neg\psi \vdash \neg\varphi$.

As in [21], we call the smallest intro-elim logic *fundamental logic*, denoted $\vdash_{\mathsf{F}}$.

Orthologic [17] is obtained from fundamental logic by adding

- *double negation elimination*: $\neg\neg\varphi \vdash \varphi$.

Compatibility logic [12] in the $\{\wedge, \vee, \neg\}$-fragment is obtained from fundamental logic by strengthening proof-by-cases in Definition 2.10 to

- *proof-by-cases with side assumptions*: if $\alpha \wedge \varphi \vdash \chi$ and $\alpha \wedge \psi \vdash \chi$, then $\alpha \wedge (\varphi \vee \psi) \vdash \psi$.

Intuitionistic logic in the $\{\wedge, \vee, \neg\}$-fragment [29] is obtained from compatibility logic by strengthening Definition 2.6/11 to the following, where $\bot$ abbreviates a contradiction such as $p \wedge \neg p$:

- *psuedocomplementation*: if $\varphi \wedge \psi \vdash \bot$, then $\psi \vdash \neg\varphi$,

Classical logic is obtained by strengthening orthologic with either proof by cases with side assumptions or pseudocomplementation (see [24, Proposition 3.7]).

In the context of a language with epistemic modals $\Diamond$ ('might') and $\Box$ ('must'), Holliday and Mandelkern [24] argue that proof-by-cases with side assumptions and pseudocomplementation are both invalid. They start by accepting well-known arguments (see Sect. 2 of [24] for a review of the relevant data) that sentences of the form 'It's raining and it might not be raining' ($p \wedge \Diamond\neg p$) are semantically contradictory, not just pragmatically infelicitous, because they embed under other operators in the way that outright contradictions ('It's raining and it's not raining') do, rather than how consistent but infelicitous Moore sentence ('It's raining but I don't know it') do. Thus, if $\Diamond$ is the epistemic modal 'might', we have $p \wedge \Diamond\neg p \vdash \bot$.[4] But then pseudocomplementation would yield

[4] Of course if we consider an alethic or counterfactual possibility modal M such as 'it could have been the case that', then $p \wedge M\neg p \nvdash \bot$.

$\Diamond\neg p \vdash \neg p$, which is absurd, since we cannot validly reason from 'It might not be raining' to 'It's not raining'. And proof-by-cases with side assumptions would yield $\Diamond p \wedge (p \vee \neg p) \vdash p$ (since $\Diamond p \wedge p \vdash p$ and $\Diamond p \wedge \neg p \vdash \bot \vdash p$), which is similarly absurd, since we cannot validly reason from 'It might be raining' together with 'It's raining or it's not raining' to 'It *is* raining'.[5]

Although we do not have the epistemic modals $\Diamond$ and $\Box$ in our formal language $\mathcal{L}$, we intend the elements of Prop from which $\mathcal{L}$ is generated to be genuine *propositional variables*, standing in for arbitrary propositions (cf. [5, pp. 147-8]). Thus, if we accept the counterexamples to pseudocomplementation and proof-by-cases with side assumptions above, then we cannot accept these principles for $\mathcal{L}$. This suggests orthologic as an appropriate base logic, as argued in [24]. But when we consider reasoning not only with epistemic modals but also with vague predicates, might this suggest going still weaker than orthologic? This is one of the questions we will consider in what follows.

3 Relational Semantics for Ortho and Fundamental Logic

To give semantics for the logics introduced in § 2, we begin with a familiar starting point from modal logic, namely relational frames.

Definition 3. A *relational frame* is a pair $(X, \lhd)$ where X is a nonempty set and $\lhd$ is a binary relation on X. A *relational model* for $\mathcal{L}$ is a triple $\mathcal{M} = (X, \lhd, V)$ where $(X, \lhd)$ is a relational frame and V assigns to each $p \in \mathsf{Prop}$ a set $V(p) \subseteq X$.

In possible world semantics for intuitionistic logic, $\lhd$ would be a preorder $\leqslant$, and $V(p)$ would be required to be a *downset*, i.e., if $x \in V(p)$ and $x' \leqslant x$, then $x' \in V(p)$. States in X are understood as information states, and $x' \leqslant x$ means that x' contains all the information that x does and possibly more.[6]

We will give a different interpretation of $\lhd$ that motivates a different constraint on $V(p)$. We read $x \lhd y$ as x *is open to* y. Paraphrasing Remark 4.2 of [21], we understand openness as follows:

- There is a distinction between *accepting* a proposition and *rejecting* it.
- It should be possible for a partial state to be completely noncommittal about a proposition, so non-acceptance of A should not entail rejection of A.
- It should be possible for a state to reject a proposition without accepting its negation; e.g., an intuitionist might reject an instance of the law of excluded middle but will not accept its negation, which is a contradiction (see Field's [11] separation of rejection, non-acceptance, and acceptance of the negation).

[5] Note that this argument does not assume that one accepts the premise 'It's raining or it's not raining' as a result of a general acceptance of excluded middle. The point is just that accepting this as a meteorological fact on a particular occasion does not license the inference from 'It might be raining' to 'It is raining'.

[6] Many authors work instead with upsets and take $x' \geqslant x$ to mean that x' contains all the information that x does and possibly more. Our formulation in terms of downsets follows, e.g., Dragalin [6].

- We say that x is *open to* y iff x does not reject any proposition that y accepts.

This picture motivates the following definition.

Definition 4. Given a relational frame $(X, \lhd)$ and $A \subseteq X$, we say that

1. x *accepts* A if $x \in A$;
2. x *rejects* A if for all $y \rhd x$, $y \notin A$;
3. x *accepts the negation of* A if for all $y \lhd x$, $y \notin A$.

Now we can state the constraint on $V(p)$ that we will use instead of the downset constraint:

if x does not accept $V(p)$, then there is a state x' open to x that rejects $V(p)$,

or formally,

$$\text{if } x \notin V(p), \text{ then } \exists x' \lhd x\, \forall x'' \rhd x'\; x'' \notin V(p). \tag{2}$$

Definition 5. We say a relational model $\mathcal{M} = (X, \lhd, V)$ is a *relational fixpoint model* if for all $p \in \mathsf{Prop}$, the set $V(p)$ satisfies (2) for all $x \in X$.

The motivation for the 'fixpoint' terminology is the following. Given a frame $(X, \lhd)$, we define a closure operator[7] $c_{\lhd} \colon \wp(X) \to \wp(X)$ by

$$c_{\lhd}(A) = \{x \in X \mid \forall x' \lhd x\, \exists x'' \rhd x' \colon x'' \in A\}.$$

A *fixpoint* of $c_{\lhd}$ is a set $A \subseteq X$ such that $c_{\lhd}(A) = A$. Since $A \subseteq c_{\lhd}(A)$ for any $A \subseteq X$, that A is a fixpoint is equivalent to $c_{\lhd}(A) \subseteq A$. The contrapositive of this is just (2) for A in place of $V(p)$:

$$\text{if } x \notin A, \text{ then } \exists x' \lhd x\, \forall x'' \rhd x' \colon x'' \notin A.$$

If we define a *proposition* in $(X, \lhd)$ to be a fixpoint of $c_{\lhd}$, we have the following.

Lemma 1. In a relational frame $(X, \lhd)$, $x \lhd y$ iff x does not reject any proposition that y accepts.

Proof. We repeat the proof from Footnote 14 of [21]. If $x \lhd y$ and y accepts A, so $y \in A$, then x does not reject A by Definition 4.2. Conversely, suppose $x \ntriangleleft y$ Now y accepts $c_{\lhd}(\{y\})$, which is a fixpoint of $c_{\lhd}$. But we claim x rejects $c_{\lhd}(\{y\})$. For if there is some z such that $x \lhd z \in c_{\lhd}(\{y\})$, then by definition of $c_{\lhd}$, there is an $x' \rhd x$ such that $x' \in \{y\}$, so $x' = y$. But this contradicts $x \ntriangleleft y$. Hence there is no z such that $x \lhd z \in c_{\lhd}(\{y\})$, which means that x rejects $c_{\lhd}(\{y\})$. □

Finally, we define the semantics of the connectives. The semantics of $\neg$ and $\wedge$ look the same as in relational semantics for intuitionistic logic (though again $\lhd$ need not be a preorder), but the semantics of $\vee$ is different. Instead of interpreting $\vee$ as the union of propositions, as in intuitionistic logic, we interpret $\vee$ as the *closure* of the union, i.e., by applying $c_{\lhd}$ to the union.

---

[7] A function $c : \wp(X) \to \wp(X)$ is a *closure operator* if for all $A, B \subseteq X$, we have $A \subseteq c(A)$, $c(c(A)) = c(A)$, and $A \subseteq B$ implies $c(A) \subseteq c(B)$.

Definition 6. Given a relational fixpoint model $\mathcal{M} = (X, \lhd, V)$ for $\mathcal{L}$, $x \in X$, and $\varphi \in \mathcal{L}$, we define $\mathcal{M}, x \Vdash \varphi$ as follows:

1. $\mathcal{M}, x \Vdash p$ iff $x \in V(p)$;
2. $\mathcal{M}, x \Vdash \neg\varphi$ iff $\forall x' \lhd x$, $\mathcal{M}, x' \nVdash \varphi$;
3. $\mathcal{M}, x \Vdash \varphi \wedge \psi$ iff $\mathcal{M}, x \Vdash \varphi$ and $\mathcal{M}, x \Vdash \psi$;
4. $\mathcal{M}, x \Vdash \varphi \vee \psi$ iff $\forall x' \lhd x\ \exists x'' \rhd x'$: $\mathcal{M}, x'' \Vdash \varphi$ or $\mathcal{M}, x'' \Vdash \psi$.

Let $[\![\varphi]\!]^{\mathcal{M}} = \{x \in X \mid \mathcal{M}, x \Vdash \varphi\}$.

The idea behind the clause for $\vee$ is that in order for x to accept $\varphi \vee \psi$, it is not required that x already accepts one of the disjuncts—instead what is required is that *no state open to x rejects both disjuncts.* (Although it may seem that we hereby take a stand against intuitionistic logic on $\vee$, see Theorem 3.1.)

We can prove by an easy induction on the structure of formulas that the semantic value of every formula is a proposition in the following sense.

Lemma 2. Let $\mathcal{M} = (X, \lhd, V)$ be a relational fixpoint model. Then for any $\varphi \in \mathcal{L}$, the set $[\![\varphi]\!]^{\mathcal{M}}$ is a fixpoint of $c_{\lhd}$.

A key feature of the states in a relational fixpoint model is that they can be *partial*—there may be some propositions that they neither accept nor reject. Thus, we can consider how the set of propositions accepted by one state relates to that of another state. For this we have a key definition and lemma.

Definition 7. Given a relational fixpoint model $\mathcal{M} = (X, \lhd, V)$ and $x, y \in X$, we say that x *pre-refines* y if for every $z \lhd x$, we have $z \lhd y$.

In symmetric models, we simply say *refines* instead of *pre-refines.* An easy induction on the structure of formulas establishes the following.

Lemma 3. Given a relational fixpoint model $\mathcal{M} = (X, \lhd, V)$ and $x, y \in X$, if x pre-refines y, then for any $\varphi \in \mathcal{L}$, $\mathcal{M}, y \Vdash \varphi$ implies $\mathcal{M}, x \Vdash \varphi$.

The final piece of setup we need are the definition of consequence over a class of models, as well as of soundness and completeness of a logic with respect to the class, all of which is completely standard.

Definition 8. Given a class C of relational models and $\varphi, \psi \in \mathcal{L}$, we define $\varphi \vDash_{\mathsf{C}} \psi$ iff for all $\mathcal{M} \in \mathsf{C}$ and states x in $\mathcal{M}$, if $\mathcal{M}, x \Vdash \varphi$, then $\mathcal{M}, x \Vdash \psi$.

We say that an intro-elim logic $\vdash_{\mathsf{L}}$ is *sound* (resp. *complete*) *with respect to* C if for all $\varphi, \psi \in \mathcal{L}$, $\varphi \vdash_{\mathsf{L}} \psi$ (resp. $\varphi \vDash_{\mathsf{C}} \psi$) implies $\varphi \vDash_{\mathsf{C}} \psi$ (resp. $\varphi \vdash_{\mathsf{L}} \psi$).

We are now prepared to state soundness and completeness theorems for four of the logics introduced in § 2. We begin with orthologic and fundamental logic.

Theorem 1 ([17])**.** Orthologic is sound and complete with respect to the class of relational fixpoint models in which $\lhd$ is reflexive and *symmetric.*

Theorem 2 ([21]). Fundamental logic is sound and complete with respect to the class of relational fixpoint models in which $\lhd$ is reflexive and *pseudosymmetric*: if $y \lhd x$, then there is a $z \lhd y$ that pre-refines x.

The key property of pseudosymmetry of $\lhd$ for fundamental logic is equivalent to the intuitive condition that *if x accepts a proposition $\neg A$, then x rejects A* (see [21, § 4.1]). Strengthening pseudosymmetry to symmetry implies the non-constructive condition that if x accepts $\neg\neg A$, then x accepts A.

As for intuitionistic and classical logic, it is easy to see that standard semantics for these logics are simply special cases of the relational fixpoint semantics (see [20, § 2] for further explanation), so from well-known completeness theorems for the logics with respect to their standard semantics, we obtain the following.

Theorem 3

1. (Possible world semantics for intuitionistic logic) Intuitionistic logic is sound and complete with respect to the class of relational fixpoint models in which $\lhd$ is reflexive and *transitive*.
2. (Possibility semantics for classical logic) Classical logic is sound and complete with respect to the class of relational fixpoint models in which $\lhd$ is reflexive, symmetric, and *compossible* [20]: whenever $x \lhd y$, there is some z that refines both x and y.
3. (Possible world semantics for classical logic) Classical logic is sound and complete with respect to the class of relational fixpoint models in which $\lhd$ is the identity relation (and $|X| = 1$).

4 Relational Semantics for Compatibility Logic

To give semantics for his *compatibility logic*, Fine [12] does not use the relational *fixpoint* models of § 3.[8] For he imposes no constraints on the valuation function V (and hence no constraints on what sets of states count as propositions) and uses a different interpretation of $\vee$, as follows.

Definition 9. Given a relational model $\mathcal{M} = (X, \lhd, V)$ for $\mathcal{L}$, $x \in X$, and $\varphi \in \mathcal{L}$, we define $\mathcal{M}, x \vDash \varphi$ as follows:

1. $\mathcal{M}, x \vDash p$ iff $x \in V(p)$;
2. $\mathcal{M}, x \vDash \neg\varphi$ iff $\forall x' \lhd x$, $\mathcal{M}, x' \nvDash \varphi$;
3. $\mathcal{M}, x \vDash \varphi \wedge \psi$ iff $\mathcal{M}, x \vDash \varphi$ and $\mathcal{M}, x \vDash \psi$;
4. $\mathcal{M}, x \vDash \varphi \vee \psi$ iff $\mathcal{M}, x \vDash \varphi$ or $\mathcal{M}, x \vDash \psi$.

Let $\|\varphi\|^{\mathcal{M}} = \{x \in X \mid \mathcal{M}, x \vDash \varphi\}$.

[8] It is possible to give a semantics for compatibility logic using fixpoint models, using results from [27, § 3.1] and [19,21], but we will not go into the details here.

The soundness of compatibility logic with respect to this semantics is easy to check. Note, in particular, from the clauses for $\wedge$ and $\vee$ that this semantics validates the distributive laws, or equivalently, proof-by-cases with side assumptions, which are invalid according to orthologic and fundamental logic. However, if we ignore disjunction, then of course Fine's semantic clauses are the same as those in §3. An obvious induction yields the following.

Lemma 4. For any relational fixpoint model $\mathcal{M} = (X, \lhd, V)$ and $\varphi \in \mathcal{L}$, if φ does not contain $\vee$, then $[\![\varphi]\!]^{\mathcal{M}} = \|\varphi\|^{\mathcal{M}}$.

Now Fine does not work with *arbitrary* relational models. He assumes the relation $\lhd$ is reflexive and symmetric. Under these assumptions, we will write '$\between$' instead of '$\lhd$' to emphasize the symmetry. Reflexivity and symmetry are the same conditions on the relation as for orthologic in Theorem 1, but the key difference is again that Fine puts no constraints on the valuation function V, so his semantics does not validate orthologic even in the $\vee$-free fragment. In particular, it does not validate the inference from $\neg\neg\varphi$ to φ, which othologic does.

We will sketch the proof that compatibility logic is complete with respect to the class of relational models in which the relation is reflexive and symmetric. Those familiar with algebraic logic will have no trouble filling in all the details. To state the key result, let us say that a unary operation $\neg$ on a bounded lattice L is a *weak pseudocomplementation* (terminology from [2,8,9]) if for all $a, b \in L$, we have that $a \wedge \neg a = 0$, that $a \leq \neg\neg a$, and that $a \leq b$ implies $\neg b \leq \neg a$.

Theorem 4. For any bounded distributive lattice L equipped with a weak pseudocomplementation $\neg$, there is a set X and a reflexive and symmetric binary relation $\between$ on X such that $(L, \neg)$ embeds into $(\wp(X), \neg_{\between})$, where $\neg_{\between} A = \{x \in X \mid \text{for all } y \between x,\ y \notin A\}$.

Proof (Sketch). Apply Stone's [30] representation of distributive lattices using prime filters to L. Then given prime filters F, F', say that $F \between F'$ if there is no $a \in L$ such that $a \in F$ and $\neg a \in F'$. □

Theorem 5. Compatibility logic is sound and complete with respect to reflexive and symmetric models under the semantics of Definition 9.

Proof (Sketch). As noted above, soundness is an easy check. For completeness, apply Theorem 4 to the Lindenbaum-Tarski algebra of compatibility logic. □

Weak pseudocomplementations are precisely the types of negations in algebras for fundamental logic (see [21, § 3]), only the underlying lattices of these algebras are not necessarily distributive. However, if we drop $\vee$ from the language, then we cannot notice this difference between compatibility logic and fundamental logic. The following theorem is essentially Proposition 4.34 of [21] but rephrased in terms of compatibility logic instead of the modal logic **KTB**.

Proposition 1. For any $\varphi, \psi \in \mathcal{L}$ not containing $\vee$, we have $\varphi \vdash \psi$ in compatibility logic if and only if $\varphi \vdash \psi$ in fundamental logic.

However, the analogous result does not hold for formulas not containing $\wedge$: as Guillaume Massas (personal communication) pointed out,

$$\neg(\neg\varphi \vee \neg(\psi \vee \chi)) \vdash \neg(\neg\varphi \vee \neg\psi) \vee \neg(\neg\varphi \vee \neg\chi)$$

in compatibility logic but not in fundamental logic, as a consequence of distributivity in the former but not the latter.

Having completed our brief tour of the logics and semantics of interest, we turn in the next section back to vagueness.

5 The Symmetric Sorites Model

Recall the Sorites Paradox from § 1. In this section, we define a relational fixpoint model for each Sorites series. The construction is inspired by a passage of Fine [12, p. 42]:

> [T]hink of the points of a model as corresponding to different admissible uses of the language. The reflexive and symmetric relation will then relate two admissible uses when they are *compatible* in the sense that there is no conflict in what is true under the one use and what is true under the other. Suppose, for example, that there are 100 men in our sorites series. Then the use in which the first 30 men are taken to be bald and the last 50 men are taken to be not bald will be compatible with the use in which the first 31 men are taken to be bald and the last 49 are taken not to be bald.

First, let us specialize our propositional language for the Sorites paradox in particular. For convenience, we now take the first item in the Sorites series to be labeled by 0 instead of 1, and we adopt the identification $n = \{0, \dots, n-1\}$.

Definition 10. For $n \in \mathbb{N}$, let $\mathcal{L}_n$ be defined like $\mathcal{L}$ in Definition 1 with $\mathsf{Prop} = \{p_k \mid k \in n\}$.

We take p_k to mean that the k-th member of the Sorites series satisfies the relevant vague predicate (e.g., 'young', 'bald', etc.).

Definition 11. For $n, \delta \in \mathbb{N}$ with $1 \leq \delta < n-1$, we define the *symmetric Sorites model* $\mathcal{S}_{n,\delta} = (S, \between, V)$ for $\mathcal{L}_n$ as follows:

1. S is the set of all pairs (i, j) for $i, j \in n \cup \{-\infty, \infty\}$ such that $i + \delta < j$;[9]
2. $(i, j) \between (i', j')$ iff $\max(i, i') < \min(j, j')$;
3. $V(p_k) = \{(i, j) \in S \mid k \leq i\}$.

Figure 1 shows the symmetric Sorites model for $n = 4$ and $\delta = 1$.

The intuition behind Definition 11 is as follows: state (i, j) will make p_k true for all $k \leq i$, by part 3 of the definition, and will make $\neg p_k$ true for all $k \geq j$ by

[9] Note that $-\infty + \delta = -\infty$.

Fact 2 below. Thus, part 2 says that two states are compatible when there is no p_k such that one state makes p_k true and the other makes $\neg p_k$ true. Finally, note that part 1 rules out sharp cutoffs: not only can a state not make p_k true and $\neg p_{k+1}$ true, but if we pick a δ greater than 1, then we can even rule out a state making p_k true and $\neg p_{k+2}$ true, etc. For example, we may not want to admit a state according to which a person who is k seconds old is young but a person who is $k + 2$ seconds old is not young. No doubt the appropriate choice of δ in a given context is itself vague. Also note the role of $-\infty$ and ∞ in part 1: the state $(-\infty, \infty)$ will not make true any formulas of the form p_k or $\neg p_k$; a state such as $(-\infty, j)$ will not make true any formulas of the form p_k; and a state such as (j, ∞) will not make true any formulas of the form $\neg p_k$.

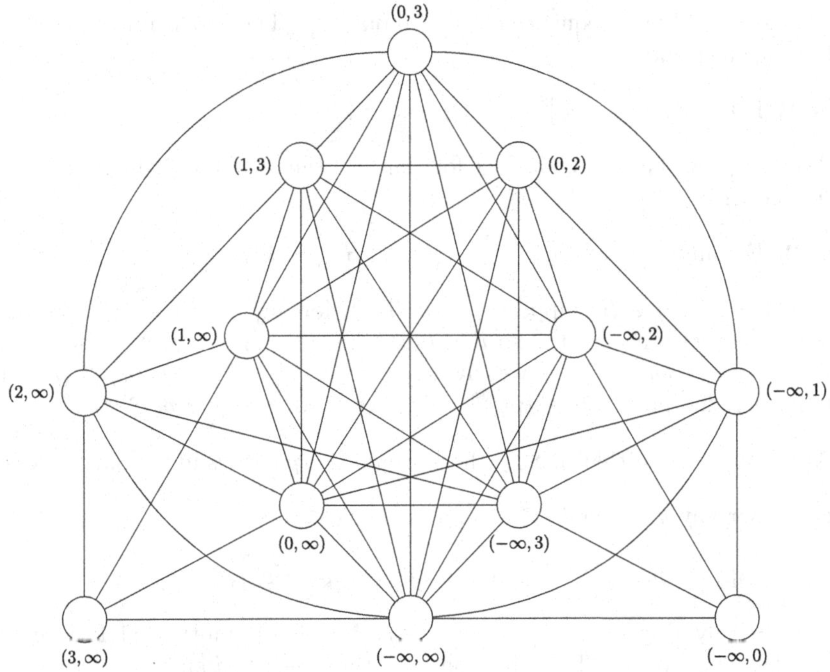

Fig. 1. The symmetric Sorites model for $n = 4$ and $\delta = 1$.

A key point about the symmetric Sorites model is that it is a model not only for Fine's semantics from § 4 but also for the fixpoint semantics from § 3.

Proposition 2. $\mathcal{S}_{n,\delta}$ is a relational fixpoint model.

Proof. We must show that $V(p_k)$ is a fixpoint of $c_{\between}$. Given $(i, j) \in S$, we have $i < j$ by Definition 11.1. Now suppose $(i, j) \notin V(p_k)$, so $i < k$ by Definition 11.3. Let $(i', j') = (-\infty, k)$, so $(i', j') \in S$ by Definition 11.1. Then $\max(i, i') = i < \min(j, j')$, so $(i', j') \between (i, j)$ by Definition 11.2. Now for any $(i'', j'') \between (i', j')$, we have $i'' < j' = k$ by Definition 11.2, so $(i'', j'') \notin V(p_k)$ by Definition 11.3. Thus, we have shown there is an $(i', j') \between (i, j)$ such that for all $(i'', j'') \between (i', j')$, we have $(i'', j'') \notin V(p_k)$, which shows that $V(p_k)$ is a fixpoint of $c_{\between}$. □

The upshot of Proposition 2 is that $\mathcal{S}_{n,\delta}$ is a model for orthologic.

Corollary 1. Since the relation $\between$ in the relational fixpoint model $\mathcal{S}_{n,\delta} = (S, \between, V)$ is reflexive and symmetric, orthologic is sound with respect to $\mathcal{S}_{n,\delta}$ according to the semantics of Definition 6, while compatibility logic is sound with respect to $\mathcal{S}_{n,\delta}$ according to the semantics of Definition 9.

Proof. That $\between$ is reflexive and symmetric is immediate from Definition 11.2. Then apply Theorems 1 and 5. □

We now build up a sequence of facts about $\mathcal{S}_{n,\delta}$. The first is immediate from Definitions 6. 1 and 11.3.

Fact 1. For any $k \in n$, $[\![p_k]\!]^{\mathcal{S}_{n,\delta}} = \{(i, j) \in S \mid k \leq i\}$.

We can prove an analogous fact for $\neg p_k$ and the *second* coordinate of states in the model.

Fact 2. For any $k \in n$, $[\![\neg p_k]\!]^{\mathcal{S}_{n,\delta}} = \{(i, j) \in S \mid j \leq k\}$.

Proof. Given $(i, j) \in S$, suppose $j \leq k$. Then for any $(i', j') \between (i, j)$, we have $i' < j \leq k$ by Definition 11.2, so $i' < k$ and hence $(i', j') \notin [\![p_k]\!]^{\mathcal{S}_{n,\delta}}$ by Fact 1. Thus, $(i, j) \in [\![\neg p_k]\!]^{\mathcal{S}_{n,\delta}}$. Conversely, suppose $k < j$. Then $(k, \infty) \between (i, j)$ by Definition 11.2, and $(k, \infty) \in [\![p_k]\!]^{\mathcal{S}_{n,\delta}}$ by Fact 1, so $(i, j) \notin [\![\neg p_k]\!]^{\mathcal{S}_{n,\delta}}$. □

We can now show that in $\mathcal{S}_{n,\delta}$, there are *no sharp cutoffs* in the Sorites series.

Fact 3. For any $k \in n$ and $\ell \in \delta + 1$ with $k + \ell \leq n - 1$,

$$[\![p_k \wedge \neg p_{k+\ell}]\!]^{\mathcal{S}_{n,\delta}} = \varnothing \text{ and hence } [\![\neg(p_k \wedge \neg p_{k+\ell})]\!]^{\mathcal{S}_{n,\delta}} = S\ .$$

Proof. For any $(i, j) \in S$, we have $i + \delta < j$ by Definition 11.1. Hence if $(i, j) \in [\![p_k]\!]^{\mathcal{S}_{n,\delta}}$, so $k \leq i$ by Definition 11.3, then $k + \delta < j$ and hence $k + \ell < j$, so $(i, j) \notin [\![\neg p_{k+\ell}]\!]^{\mathcal{S}_{n,\delta}}$ by Fact 2. □

We can now put everything together to show the joint satisfiability of the claims that (i) the first element of the Sorites series satisfies the relevant predicate, (ii) the last element does not, and (iii) there are no sharp cutoffs.

Fact 4. For any $n, \delta \in \mathbb{N}$ with $1 \leq \delta < n - 1$, we have

$$\mathcal{S}_{n,\delta}, (0, n-1) \Vdash p_0 \wedge \neg p_{n-1} \wedge \bigwedge_{0 \leq k \leq n-2} \ \bigwedge_{\substack{1 \leq \ell \leq \delta \\ k+\ell \leq n-1}} \neg(p_k \wedge \neg p_{k+\ell}).$$

Proof. Since $\delta < n-1$, we have $(0, n-1) \in S$ by Definition 11.1. Then we have $\mathcal{S}_{n,\delta}, (0, n-1) \Vdash p_0$ by Fact 1; $\mathcal{S}_{n,\delta}, (0, n-1) \Vdash \neg p_{n-1}$ by Fact 2; and finally $\mathcal{S}_{n,\delta}, (0, n-1) \Vdash \bigwedge_{0 \leq k \leq n-2} \bigwedge_{\substack{1 \leq \ell \leq \delta \\ k+\ell \leq n-1}} \neg(p_k \wedge \neg p_{k+\ell})$ by Fact 3. □

Fact 5. For any $n, \delta \in \mathbb{N}$ with $1 \leq \delta < n-1$, the formula

$$p_0 \wedge \neg p_{n-1} \wedge \bigwedge_{0 \leq k \leq n-2} \bigwedge_{\substack{1 \leq \ell \leq \delta \\ k+\ell \leq n-1}} \neg(p_k \wedge \neg p_{k+\ell})$$

is consistent in orthologic and compatibility logic.

Proof. Immediate from Fact 4, Lemma 4, and Corollary 1. □

The significance of Fact 5 is that we can consistently deny the existence of sharp cutoffs in a Sorites series by either (a) denying distributivity, while accepting excluded middle, as in orthologic or (b) denying excluded middle, while accepting distributivity, as in compatibility logic. Which approach is preferable? If we endorse the arguments concerning epistemic modals mentioned in § 1, then accepting distributivity as a generally valid principle is not an available path. Still, we are left with the question of whether to accept excluded middle.

Before turning to that question in § 6, let us address another natural question: is there something in common between the case of epistemic modals and the case of vague predicates that explains why an orthological treatment might be appropriate for both? The semantics used in this section provides a hint. Both epistemic modals and vagueness are related to *partial states*, whether due to the partiality of information or the partiality of predicate determination. Moreover, these states arguably violate a key assumption underlying classical semantics identified in [23,24]: that *compatibility implies compossibility*, where two states are compossible if they have a common refinement (recall Theorem 3.2). In the case of epistemic modals, a state satisfying $\Diamond\neg p$ can be compatible with a state satisfying p, since $\Diamond\neg p$ does not entail $\neg p$, but they cannot have a common refinement, since such a refinement would satisfy $p \wedge \Diamond\neg p$, an epistemic contradiction (recall § 2). In the case of vagueness, in particular in the model of the Sorites in Fig. 1, a state satisfying p_1 can be compatible with a state satisfying $\neg p_2$ (e.g., $(1,3)$ is compatible with $(0,2)$), since p_1 does not entail p_2, but they cannot have a common refinement, since such a refinement would satisfy $p_1 \wedge \neg p_2$, contradicting our prohibition against sharp cutoffs in the Sorites. Thus, the cases of epistemic modals and vagueness are tied together by the relevant states violating the classical dictum that compatibility implies compossibility.

6 A Pseudosymmetric Sorites Model

In this section, working with the fixpoint semantics of § 3, we add to the symmetric Sorites model the possibility of *rejecting* $p_k \vee \neg p_k$, for each $k \in n$. Following Definition 4, we say that a state x rejects a formula φ if for all $y \rhd x$, we have

$\mathcal{M}, y \nVdash \varphi$. In symmetric models, this is equivalent to x forcing $\neg\varphi$, but in pseudosymmetric models, it is not. In particular, in pseudosymmetric models, it is possible for a state to reject an instance of excluded middle, as intuitionists may do. In the following construction, k will be the state that rejects $p_k \vee \neg p_k$.[10]

Definition 12. For $n, \delta \in \mathbb{N}$ with $1 \leq \delta < n - 1$, where $\mathcal{S}_{n,\delta} = (S, \between, V)$ is the symmetric Sorites model for $\mathcal{L}_n$ as in Definition 11, we define the *pseudosymmetric Sorites model* $\mathfrak{S}_{n,\delta} = (X, \lhd, V)$ as follows:

1. $X = S \cup n$;
2. if $(i, j), (i', j') \in S$, then $(i, j) \lhd (i', j')$ iff $(i, j) \between (i', j')$;
3. if $k \in n$ and $(i, j) \in S$, then $k \lhd (i, j)$ iff $i < k < j$;
4. if $x \in X$ and $k \in n$, then $x \lhd k$.

Note that the valuation V in $\mathfrak{S}_{n,\delta}$ is the same as in $\mathcal{S}_{n,\delta}$.

The pseudosymmetric Sorites model for $n = 4$ and $\delta = 1$ is shown in Fig. 2.

Proposition 3. $\mathfrak{S}_{n,\delta}$ is a fixpoint model.

Proof. We adapt the proof of Proposition 2 to show that $V(p_k)$ is a fixpoint of $c_\lhd$. Suppose $x \notin V(p_k)$.

Case 1: $x = (i, j) \in S$. Hence $i < j$ by Definition 11.1. Since $(i, j) \notin V(p_k)$, we have $i < k$ by Definition 11.3. Let $(i', j') = (-\infty, k)$, so $(i', j') \in S$ by Definition 11.1. Then $\max(i, i') = i < \min(j, j')$, so $(i', j') \between (i, j)$ by Definition 11.2 and hence $(i', j') \lhd (i, j)$ by Definition 12.2. Now consider any $y \rhd (i', j')$. If $y = (i'', j'') \in S$, then $(i'', j'') \rhd (i', j')$ implies $(i'', j'') \between (i', j')$ by Definition 12.2, so we have $i'' < j' = k$ by Definition 11.2, so $(i'', j'') \notin V(p_k)$ by Definition 11. 3. On the other hand, if $y = \ell \in n$, then $\ell \notin V(p_k)$ since none of the new states from n are in any $V(p_k)$. Thus, we have shown there is an $(i', j') \lhd (i, j)$ such that for all $y \rhd (i', j')$, we have $y \notin V(p_k)$.

Case 2: $x = \ell \in n$. Then again we take $(i', k') = (-\infty, k)$, so $(i', k') \lhd \ell$ by Definition 12.4. Then the rest of the argument is as in Case 1.

In either case, there is an $(i', j') \lhd x$ such that for all $y \rhd (i', j')$, we have $y \notin V(p_k)$. This shows that $V(p_k)$ is a fixpoint of $c_\lhd$. □

Though the addition of the new states in n breaks the symmetry of $\lhd$ in $\mathfrak{S}_{n,\delta}$, we retain the property of pseudosymmetry needed for fundamental logic.

Proposition 4. In $\mathfrak{S}_{n,\delta} = (X, \lhd, V)$, $\lhd$ is reflexive and pseudosymmetric, so fundamental logic is sound with respect to $\mathfrak{S}_{n,\delta}$ according to the semantics of Definition 6.

Proof. Reflexivity is obvious. For pseudosymmetry, by Definition 12.2, the restriction of $\lhd$ to S is symmetric; by Definition 12.4, the restriction of $\lhd$ to n is symmetric; and by Definition 12.4, $k \lhd (i, j)$ implies $(i, j) \lhd k$. Thus, we need only consider the case where $(i, j) \lhd k$. In this case, we simply observe that $(i, j) \rhd (i, j)$ by Definitions 12.2 and 11.2 , and (i, j) pre-refines k by Definition 12.4 . This establishes pseudosymmetry. □

[10] One could also add, for each interval $I \subseteq n$, a state x_I that rejects $p_k \vee \neg p_k$ for each $k \in I$. But it suffices to make our points here to just add one state for each p_k.

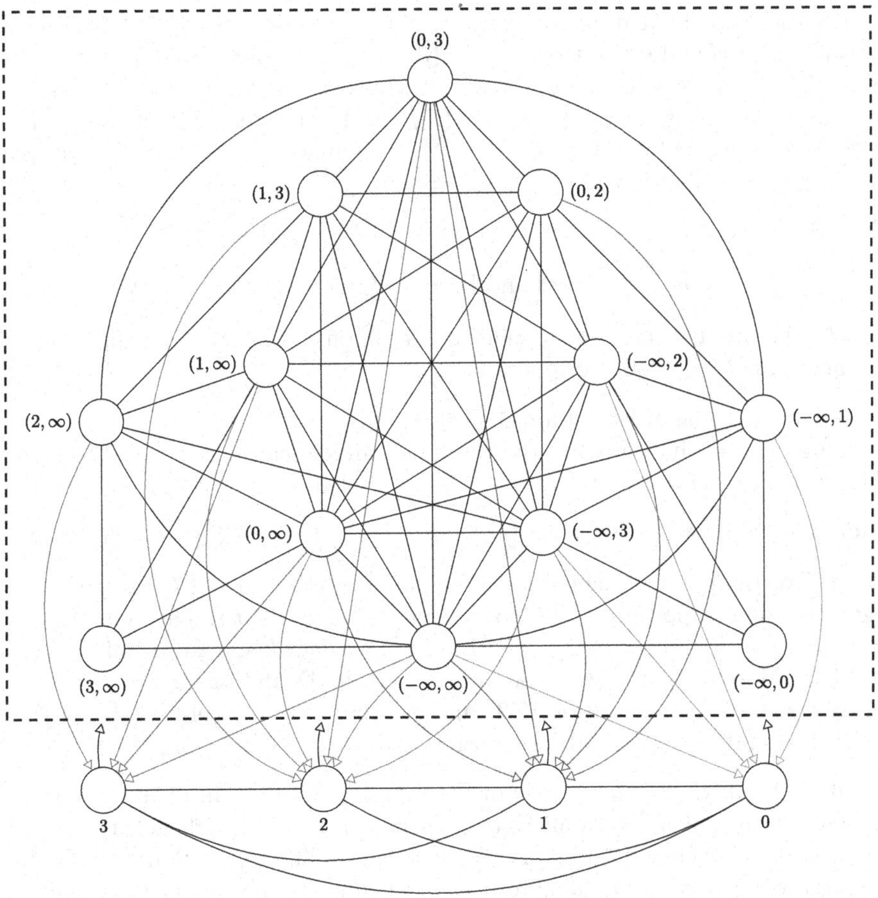

Fig. 2. The pseudosymmetric Sorites model for $n = 4$ and $\delta = 1$. We indicate $x \lhd y$ by an arrow from y to x. The arrow from k to the dashed rectangle indicates arrows from k to all states inside the rectangle. An edge with no arrow head indicates arrows in both direction.

We now prove a series of facts analogous to those for $\mathcal{S}_{n,\delta}$ in § 5.

Fact 6. For any $k \in n$, $[\![p_k]\!]^{\mathfrak{S}_{n,\delta}} = \{(i,j) \in S \mid k \leq i\}$.

Proof. Immediate from Definitions 6.1, 12, and 11.3.

Fact 7. For any $k \in n$,

$$[\![\neg p_k]\!]^{\mathfrak{S}_{n,\delta}} = \{(i,j) \in S \mid j \leq k\} \text{ and } [\![\neg\neg p_k]\!]^{\mathfrak{S}_{n,\delta}} = [\![p_k]\!]^{\mathfrak{S}_{n,\delta}} .$$

Proof. The proof of the first equation is essentially the same as that of Fact 2, only using the facts that (i) none of the new states $\ell \in n$ belong to any $V(p_k)$, and (ii) the new states $\ell \in n$ are such that $x \lhd \ell$ for all $x \in X$.

For the second equation, we have $[\![p_k]\!]^{\mathfrak{S}_{n,\delta}} = \{(i,j) \in S \mid k \leq i\}$ by Fact 6. Now if $(i,j) \in S$ and $k \leq i$, then for any $(i',j') \lhd (i,j)$, we have $k < j'$ by Definitions 12.2 and 11.2, so $(i',j') \notin [\![\neg p_k]\!]^{\mathfrak{S}_{n,\delta}}$, which shows that $(i,j) \in [\![\neg\neg p_k]\!]^{\mathfrak{S}_{n,\delta}}$. Conversely, if $i < k$, then $(-\infty, k) \lhd (i,j)$ by Definitions 12.2 and 11. 2, and $(-\infty,k) \in [\![\neg p_k]\!]^{\mathfrak{S}_{n,\delta}}$, so $(i,j) \notin [\![\neg\neg p_k]\!]^{\mathfrak{S}_{n,\delta}}$. Similarly, for each $k \in n$, we have $(-\infty,k) \lhd$ by Definition 12.4, so $k \notin [\![\neg\neg p_k]\!]^{\mathfrak{S}_{n,\delta}}$. □

Fact 8. For any $k \in n$ and $\ell \in \delta + 1$ with $k + \ell \leq n - 1$,

$$[\![p_k \wedge \neg p_{k+\ell}]\!]^{\mathfrak{S}_{n,\delta}} = \varnothing \text{ and hence } [\![\neg(p_k \wedge \neg p_{k+\ell})]\!]^{\mathfrak{S}_{n,\delta}} = X\ .$$

Proof. The proof is the same as that of Fact 3, only using the fact that none of the new states $k \in n$ belong to any $V(p_k)$. □

Thus, the analogue of Fact 4 for $\mathfrak{S}_{n,\delta}$ also holds.

What is new about $\mathfrak{S}_{n,\delta}$ is that excluded middle can now be used to define the set of states at which the status of p_k is settled.

Fact 9. For any $k \in n$, we have $[\![p_k \vee \neg p_k]\!]^{\mathfrak{S}_{n,\delta}} = \{(i,j) \in S \mid k \leq i \text{ or } j \leq k\}$.

Proof. The right-to-left inclusion is immediate from Facts 6 and 7. For the left-to-right inclusion, suppose $x \in X \setminus \{(i,j) \in S \mid k \leq i \text{ or } j \leq k\}$. Then either $x \in n$ or $x = (i,j) \in S$ and $i < k < j$. In either case, we have $k \lhd x$ by Definitions 12.3 and 12.4. Moreover, for any $y \rhd k$, either $y = k$ by Definition 12.4 or $y = (i',j')$ where $i' < k < j'$ by Definition 12.3 , and in either case, $y \notin [\![p_k]\!]^{\mathfrak{S}_{n,\delta}} \cup [\![\neg p_k]\!]^{\mathfrak{S}_{n,\delta}}$ by Facts 6 and 7. It follows by Definition 6.4 that $x \notin [\![p_k \vee \neg p_k]\!]^{\mathfrak{S}_{n,\delta}}$. □

In addition, $\mathfrak{S}_{n,\delta}$ exhibits one of Fine's [12, p. 44ff] main claims about the "global" nature of indeterminacy in a Sorites series: although negating a single instance of excluded middle is of course contradictory, when dealing with a Sorites series we should be able to negate a conjunction of distinct instances of excluded middle or even weak excluded middle (i.e., $\neg\varphi \vee \neg\neg\varphi$).[11]

Fact 10. In $\mathfrak{S}_{n,\delta}$, we have

$$\begin{aligned}&[\![(p_0 \vee \neg p_0) \wedge \cdots \wedge (p_{n-1} \vee \neg p_{n-1})]\!]^{\mathfrak{S}_{n,\delta}}\\ =\ &[\![(\neg p_0 \vee \neg\neg p_0) \wedge \cdots \wedge (\neg p_{n-1} \vee \neg\neg p_{n-1})]\!]^{\mathfrak{S}_{n,\delta}}\\ =\ &\{(n-1,\infty), (-\infty, n-1)\},\end{aligned}$$

so

$$\begin{aligned}&[\![\neg((p_0 \vee \neg p_0) \wedge \cdots \wedge (p_{n-1} \vee \neg p_{n-1}))]\!]^{\mathfrak{S}_{n,\delta}}\\ =\ &[\![\neg((\neg p_0 \vee \neg\neg p_0) \wedge \cdots \wedge (\neg p_{n-1} \vee \neg\neg p_{n-1}))]\!]^{\mathfrak{S}_{n,\delta}}\\ =\ &\{(i,j) \in S \mid i,j \in \mathbb{N}\}.\end{aligned}$$

[11] Fine [12, p. 34] suggests that "under fairly innocuous logical assumptions", the negation of the conjunction of instances of excluded middle is equivalent to the conjunction of negations of state descriptions of the form $p_1 \wedge \cdots \wedge p_k \wedge \neg p_{k+1} \wedge \cdots \wedge \neg p_n$. But orthologic makes the former unsatisfiable while the latter satisfiable (Fact 4), and arguably distributivity is not an innocuous logical assumption.

Proof. We already know the formulas using excluded middle and weak excluded middle have the same extension by the second part of Fact 7.

Now if $x \in [\![(p_0 \vee \neg p_0) \wedge \cdots \wedge (p_{n-1} \vee \neg p_{n-1})]\!]^{\mathfrak{S}_{n,\delta}}$, then by Fact 9, $x = (i, j) \in S$ and for every $k \in n$, $k \leq i$ or $j \leq k$. It follows that $i = n - 1$ or $j = n - 1$, which implies $x = (n - 1, \infty)$ or $x = (-\infty, n - 1)$ by Definition 11.1. For the second part, the only states y for which neither $(n - 1, \infty) \lhd y$ nor $(-\infty, n - 1) \lhd y$ are the pairs of natural numbers. Hence $[\![\neg((p_0 \vee \neg p_0) \wedge \cdots \wedge (p_{n-1} \vee \neg p_{n-1}))]\!]^{\mathfrak{S}_{n,\delta}} = \{(i, j) \in S \mid i, j \in \mathbb{N}\}$. □

7 Discussion

Let us now return to our initial question: what is a natural non-classical base logic to which to retreat in light of both the non-classicality emerging from epistemic modals and the non-classicality emerging from vagueness? We are simply taking for granted here the arguments against distributivity involving epistemic modals from [24]. As we have seen in § 5, dropping distributivity as in orthologic is also enough to render consistent the denial of sharp cutoffs in the Sorites. Thus, a temptingly economical view is that the key to dealing with both the non-classicality emerging from epistemic modals and the non-classicality emerging from vagueness is to deny distributivity and retreat to orthologic.

However, we have also seen in § 6 a potential benefit of going still weaker than orthologic, down to fundamental logic. In the fundamental approach to the Sorites, as in Fine's approach, instances of excluded middle such as $p_k \vee \neg p_k$ have genuine expressive value (recall Fact 9): rather than being empty tautologies, as in orthologic, in the fundamental approach they express that *there is a fact of the matter* about whether a given person is young or not, whether they are bald or not, etc., a function for excluded middle highlighted by Field [11]. Thus, we have the ability to express for some questions—e.g., whether there is phosphene on Venus—that there is a fact of the matter (*phosphene* $\vee$ ¬*phosphene*), while withholding such claims for other questions—e.g., whether Jerry Fallwell's life began on a particular nanosecond, to borrow Field's example. It is true that we still cannot assert $\neg(p \vee \neg p)$, which is inconsistent, but we are able to withhold assent from $p \vee \neg p$; and as Fine [12, p. 39f] stresses, we can even use negation to deny that for every item in a Sorites series, there is a fact of the matter about whether the relevant predicate applies (recall Fact 10).

From Fine's point of view, there would seem to be another objection to orthologic. In orthologic, $\neg(p_k \wedge \neg p_{k+1})$ is equivalent to $\neg p_k \vee p_{k+1}$, so orthologic must treat the conjunctive and disjunctive versions of the Sorites as equivalent. Yet Fine [13, p. 737] wants to treat them differently: for the disjunctive version, disjunctive syllogism is valid on Fine's approach, but he does not accept the Sorites premises of the form $\neg p_k \vee p_{k+1}$. By contrast, in orthologic, $\neg p_k \vee p_{k+1}$ is equivalent to $\neg(p_k \wedge \neg p_{k+1})$, which Fine accepts, but disjunctive syllogism is not valid in orthologic.[12] As in Fine's approach, in the fundamental approach of § 6, $\neg p_k \vee p_{k+1}$ is not equivalent to $\neg(p_k \wedge \neg p_{k+1})$; while the latter is forced at all states in $\mathfrak{S}_{n,\delta}$, $\neg p_0 \vee p_1$ is not forced at, e.g., the state $(0,3)$.[13] Indeed, in $\mathfrak{S}_{n,\delta}$, a state forces $\neg p_k \vee p_{k+1}$ only if it forces $\neg p_k$ or forces p_{k+1}. Thus, although disjunctive syllogism is not schematically valid in fundamental logic, in the model $\mathfrak{S}_{n,\delta}$, any state that forces $p_k \wedge (\neg p_k \vee p_{k+1})$ also forces p_{k+1}.

Although we did not include a conditional in our language in this paper, if we were to add a conditional, this might provide another argument in favor of the fundamental approach over the orthological one. Suppose we accept that the "Or-to-If" inference from $\neg p_k \vee p_{k+1}$ to $p_k \to p_{k+1}$ preserves certainty.[14] Then if we are certain there is no sharp cutoff, the orthological approach would commit us to certainty in $\neg p_k \vee p_{k+1}$ for each k, in which case Or-to-If would commit us to certainty in $p_k \to p_{k+1}$ for each k; hence we would have to deny modus ponens for $\to$ in order to block the derivation of p_n. Though there may be reasons to deny modus ponens for conditionals whose consequents contain epistemic modals or conditionals [28], being forced to deny modus ponens for simple conditionals—albeit with vague predicates—may seem more costly. Perhaps a proponent of the orthological approach to vagueness could escape this modus ponens problem by denying that we are certain that there are no sharp cutoffs, or by denying that Or-to-If preserves certainty when vague predicates are involved, inspired by Fine's rejection of the move from $\neg(p_k \wedge \neg p_{k+1})$ to $p_k \to p_{k+1}$. Whichever path one chooses, the cost-benefit analysis seems likely to be quite subtle.

Let us suppose for the moment that one is convinced by some of the considerations above to give up the orthological principles that are rejected by the intuitionists, namely excluded middle and the inference from $\neg(\varphi \wedge \neg\psi)$ to $\neg\varphi \vee \psi$. Combined with the arguments against distributivity (or proof-by-cases with side assumptions) involving epistemic modals mentioned in § 1, this would lead us toward fundamental logic. Still, one could admit that certain principles that are not generally valid are safe when restricted to propositions of special types. For example, one could hold that formulas without epistemic modals express propositions in a special subalgebra of the ambient algebra of propositions; in [24] this is a Boolean algebra, but if vague predicates are allowed, it could instead be a bounded distributive lattice with a weak pseudocomplementation, $(L, \neg)$. Then formulas containing neither epistemic modals nor vague predicates could be taken to express propositions in a Boolean subalgebra of $(L, \neg)$.

---

[12] For example, in the epistemic orthologic of [24], one can accept $p \vee \Box\neg p$ (e.g., "Either the cat is inside or he must be outside") and $\Diamond p$ ("The cat might be inside"), without concluding p ("The cat is inside") (cf. [25, citing Yalcin]).

[13] This is because $1 \lhd (0,3)$, and for any $x \rhd 1$, $x \notin [\![p_1]\!]^{\mathfrak{S}_{n,\delta}}$ and $x \notin [\![\neg p_1]\!]^{\mathfrak{S}_{n,\delta}}$. Then since $x \in [\![\neg p_0]\!]^{\mathfrak{S}_{n,\delta}}$ implies $x \in [\![\neg p_1]\!]^{\mathfrak{S}_{n,\delta}}$ in $\mathfrak{S}_{n,\delta}$, we also have $x \notin [\![\neg p_0]\!]^{\mathfrak{S}_{n,\delta}}$.

[14] Arguably the inference is not *valid*, assuming the standard view that probability is monotonic with respect to logical consequence, as shown by the fact that one can rationally assign higher probability to $\neg p \vee q$ than to $p \to q$. To use an example from Bas van Fraassen, it is rational to assign 'The die did not land on an odd number or it landed on 5' probability 4/6 and 'If the die landed on an odd number, then it landed on 5' probability 1/3.

In a sense, nothing is lost by moving from orthologic to fundamental logic. For the Gödel-Gentzen translation from classical logic to intuitionistic logic [14,16] is also a full and faithful embedding of orthologic into fundamental logic. Recall this is the translation g defined by

$$\begin{aligned} g(p) &= \neg\neg p \\ g(\neg\varphi) &= \neg g(\varphi) \\ g(\varphi \wedge \psi) &= (g(\varphi) \wedge g(\psi)) \\ g(\varphi \vee \psi) &= g(\neg(\neg\varphi \wedge \neg\psi)). \end{aligned}$$

Proposition 5 ([21])**.** For all $\varphi, \psi \in \mathcal{L}$, we have $\varphi \vdash \psi$ in orthologic if and only if $g(\varphi) \vdash g(\psi)$ in fundamental logic.

Thus, all orthological reasoning can be carried out inside fundamental logic,[15] while fundamental logic has the advantage that orthological tautologies such as excluded middle have genuine expressive value—and hence withholding assent from them also has genuine expressive value.

Might some other sublogic of orthologic and compatibility logic be appropriate for handling the non-classicality coming from epistemic modals and vagueness? Consider, for example, the *intersection* of orthologic and compatibility logic: $\varphi \vdash \psi$ in the intersection logic iff $\varphi \vdash \psi$ in both orthologic and compatibility logic. This logic is stronger than fundamental logic, as it includes principles such as $p \wedge (q \vee r) \vdash (p \vee \neg p) \vee ((p \wedge q) \vee (p \wedge r))$, since the left disjunct of the conclusion follows in orthologic and the right disjunct follows in compatibility logic. However, it is doubtful that there is a well-motivated proof theory or natural semantics for this intersection logic. By contrast, fundamental logic has a well-motivated Fitch-style proof theory based on introduction and elimination rules [21], as well as a sequent calculus [1], and a natural semantics using reflexive and pseudosymmetric frames. Until a similarly natural sublogic of orthologic and compatibility logic appears, it appears that fundamental logic is the most natural sublogic of orthologic and compatibility logic to consider for accommodating the non-classicality coming from epistemic modals and from vagueness.

8 Conclusion

It is a tantalizing possibility that the system of orthologic first discovered in the context of quantum mechanics [3] could solve both the puzzle of epistemic modals [24] and the paradox of the Sorites. Whether vagueness also provides motivation to go weaker than orthologic, perhaps down to the system of fundamental logic, is a question we have discussed but not settled. Of course, there are other motivations for such a weakening, coming from the tradition of constructive mathematics [32]. In fact, some reasons for rejecting excluded middle

[15] In fact, all classical propositional reasoning can also be carried out inside fundamental logic at the expense of an exponential blowup in the "translation"; see Proposition 2.4 of [22].

in mathematical contexts are not entirely unrelated to reasons having to do with vagueness. For example, Feferman [10] rejects the instance of excluded middle for the Continuum Hypothesis on the grounds that "the concept of the totality of arbitrary subsets of A is essentially underdetermined or vague" (p. 21).

Here we have drawn inspiration from Fine's [12] non-classical approach to vagueness. There are of course other approaches to vagueness that attempt to salvage more of classical logic, such as supervaluationist and epistemicist approaches (see [33]), or at least to hold the line at intuitionistic logic [4,34]. The orthological and fundamental approaches to vagueness should be systematically compared to these other approaches. But those who accept (1) concerning the Sorites Paradox cannot accept the classical or even intuitionistic views of $\wedge$ and $\neg$. We hope the models discussed in this paper render more intelligible the consistency of (1) according to alternative views of the connectives.

Acknowledgements. I thank the two reviewers for valuable comments, as well as Ahmee Christensen, Matt Mandelkern, and Guillaume Massas for helpful discussion. I also thank the organizers of the 4th Tsinghua Interdisciplinary Workshop on Logic, Language, and Meaning for the invitation to speak on the topic of The Connectives in Logic and Language, which prompted this paper.

References

1. Aguilera, J.P., Bydžovský, J.: Fundamental logic is decidable. ACM Trans. Comput. Logic **25**(3), 17:1–17:14 (2024). https://doi.org/10.1145/3665328
2. Almeida, A.: Canonical extensions and relational representations of lattices with negation. Stud. Logica. **91**(2), 171–199 (2009). https://doi.org/10.1007/s11225-009-9171-8
3. Birkhoff, G., von Neumann, J.: The logic of quantum mechanics. Ann. Math. **37**(4), 823–843 (1936)
4. Bobzien, S., Rumfitt, I.: Intuitionism and the Modal Logic of Vagueness. J. Philos. Log. **49**(2), 221–248 (2020). https://doi.org/10.1007/s10992-019-09507-x
5. Burgess, J.P.: Which modal models are the right ones (for logical necessity)? Theoria **18**(47), 145–158 (2003). https://doi.org/10.1387/theoria.418
6. Dragalin, A.G.: Mathematical Intuitionism: Introduction to Proof Theory, Translations of Mathematical Monographs, vol. 67. American Mathematical Society, Providence, RI (1988)
7. Dummett, M.A.E., Lemmon, E.J.: Modal logics between S4 and S5. Zeitschrift für Mathematische Logik und Grundlagen der Mathematik **5**, 250–264 (1959). https://doi.org/10.1002/malq.19590051405
8. Dzik, W., Orlowska, E., van Alten, C.: Relational representation theorems for general lattices with negations. In: Schmidt, R.A. (ed.) RelMiCS 2006. LNCS, vol. 4136, pp. 162–176. Springer, Heidelberg (2006). https://doi.org/10.1007/11828563_11
9. Dzik, W., Orlowska, E., van Alten, C.: Relational representation theorems for lattices with negations: a survey. Lect. Notes Artif. Intell. **4342**, 245–266 (2006). https://doi.org/10.1007/11964810_12
10. Feferman, S.: Is the continuum hypothesis a definite mathematical problem? (2011). https://math.stanford.edu/~feferman/papers/IsCHdefinite.pdf

11. Field, H.: No fact of the matter. Australas. J. Philos. **81**(4), 457–480 (2003). https://doi.org/10.1080/713659756
12. Fine, K.: Vagueness: A Global Approach. Oxford University Press, New York (2020)
13. Fine, K.: In defense of a global view of vagueness: response to Andreas Ditter's 'Fine on the possibility of vagueness'. In: Faroldi, F., Van De Putte, F. (eds.) Kit Fine on Truthmakers, Relevance, and Non-classical Logic, Outstanding Contributions to Logic, vol. 26. Springer, Cham (2023). https://doi.org/10.1007/978-3-031-29415-0_32
14. Gentzen, G.: Die Widerspruchsfreiheit der reinen Zahlentheorie. Mathematische Annalen **112**, 493–565 (1936). English translation in [31], pp. 132-213
15. Gödel, K.: Collected Works. Oxford University Press, New York (1986)
16. Gödel, K.: Zur intuitionistischen Arithmetik und Zahlentheorie. Ergebnisse eines Mathematischen Kolloquiums **4**, 34–38 (1933). English translation in [15], pp. 286-295
17. Goldblatt, R.I.: Semantic analysis of orthologic. J. Philos. Logic **3**(1), 19–35 (1974). https://doi.org/10.1007/BF00652069
18. Grzegorczyk, A.: A philosophically plausible formal interpretation of intuitionistic logic. Indag. Math. **26**, 596–601 (1964)
19. Holliday, W.H.: Three roads to complete lattices: orders, compatibility, polarity. Algebra Universalis **82**(2), 1–14 (2021). https://doi.org/10.1007/s00012-021-00711-y
20. Holliday, W.H.: Compatibility and accessibility: lattice representations for semantics of non-classical and modal logics. In: Duque, D.F., Palmigiano, A. (eds.) Advances in Modal Logic, Vol. 14, pp. 507–529. College Publications, London (2022). arXiv:2201.07098
21. Holliday, W.H.: A fundamental non-classical logic. Logics **1**(1), 36–79 (2023). https://doi.org/10.3390/logics1010004
22. Holliday, W.H.: Modal logic, fundamentally. In: Ciabattoni, A., Gabelaia, D., Sedlár, I. (eds.) Advances in Modal Logic, vol. 15, pp. 423–446. College Publications, London (2024). arXiv:2403.14043
23. Holliday, W.H., Mandelkern, M.: Compatibility, compossibility, and epistemic modality. In: Proceedings of the 23rd Amsterdam Colloquium, pp. 120–126 (2022)
24. Holliday, W.H., Mandelkern, M.: The orthologic of epistemic modals. J. Philos. Logic **53**, 831–907 (2024). https://doi.org/10.1007/s10992-024-09746-7, arXiv:2203.02872
25. Klinedinst, N., Rothschild, D.: Connectives without truth-tables. Nat. Lang. Seman. **20**, 137–175 (2012). https://doi.org/10.1007/s11050-011-9079-5
26. Kripke, S.A.: Semantical analysis of intuitionistic logic I. In: Crossley, J.N., Dummett, M.A.E. (eds.) Formal Systems and Recursive Functions, pp. 92–130. North-Holland, Amsterdam (1965)
27. Massas, G.: B-frame duality. Ann. Pure Appl. Logic **174**(5), 103245 (2023). https://doi.org/10.1016/j.apal.2023.103245
28. McGee, V.: A counterexample to modus ponens. J. Philos. **82**(9), 462–471 (1985). https://doi.org/10.2307/2026276
29. Rebagliato, J., Verdú, V.: On the algebraization of some Gentzen systems. Fund. Inform. **17**(2–4), 319–338 (1993). https://doi.org/10.3233/FI-1993-182-417
30. Stone, M.: Topological representation of distributive lattices and Brouwerian logics. Časopis pro pěstování matematiky a fysiky **67**(1), 1–25 (1938)
31. Szabo, M.E. (ed.): The Collected Papers of Gerhard Gentzen. North-Holland Publishing Company, Amsterdam (1969)

32. Troelstra, A.S., van Dalen, D.: Constructivism in Mathematics, vol. I. North-Holland, Amsterdam (1988)
33. Williamson, T.: Vagueness. Routledge, New York, NY (1994)
34. Wright, C.: On being in a quandary: relativism, vagueness, logical revisionism. Mind **110**(437), 45–98 (2001). https://doi.org/10.1093/mind/110.437.45

A Probabilistic Logic for Causal Counterfactuals

Jingzhi Fang(✉)

Institute of Logic and Cognition, Department of Philosophy, Sun Yat-sen University, Guangzhou, China
fangjzh5@mail.sysu.edu.cn

Abstract. The causal modelling approach has been one of the most prominent approaches to causation over recent decades. It appeals to functional causal models to characterize dependence between events represented by variables and values in causal systems. The crucial operation "intervention" in the models makes it possible to change values of variables and talk about a certain kind of counterfactuals, e.g. "if variable X were set to its value x, then variable Y would take value y." We call the counterfactuals based on the intervention operation and expressed by variables with values causal counterfactuals. Halpern provided deterministic semantics for these counterfactuals and the corresponding axiomatic systems. This paper proposes probabilistic interpretations for causal counterfactuals in probabilistic causal models defined by Pearl, and studies their logics. An application of the probabilistic semantics is to interpret counterfactuals in counterfactual theories of actual causation, so as to obtain indeterministic versions of the theories. The new probabilistic accounts can be treated as the semantics for causal conditionals, for which I explore the validity of some common conditional properties.

Keywords: Causal counterfactual · Probabilistic causal model · Causal conditional

1 Introduction

Causal and counterfactual reasoning has been an important topic in logic, linguistics and artificial intelligence research, and the recent development of *causal modelling* techniques suggests interesting directions to approach many related issues. The causal modelling approach to causality appeals to *structural equation models* or *functional causal models* to capture the core features of causal systems: a set of (exogenous and endogenous) variables and the dependencies between them. A central notion in causal models is *intervention*: a "manipulation" operation that is useful to define what we call *causal counterfactuals* or *interventionist conditionals*. For instance, the truth of a sentence "if an arsonist had not dropped a lit match ($MD = 0$), then the forest fire would not have

J. Yan et al. (Eds.): TLLM 2024, LNCS 15410, pp. 23–50, 2025.
https://doi.org/10.1007/978-3-031-86054-6_2

occurred ($FF = 0$)" or "a variable FF would take value 0 had MD been set to value 0" is evaluated in a causal model by means of the intervention of forcing MD to take its value 0. A causal counterfactual of the form $\mathbf{X} = \mathbf{x} \Box\!\!\rightarrow \varphi$, interpreted in causal models as "if a set of variables $\mathbf{X}$ were (had been) intervened to take a value setting $\mathbf{x}$, φ would be (have been) true." The main purpose of this work is to provide a probabilistic interpretation for such causal counterfactuals in probabilistic causal models defined by Pearl [29, pp. 205-6] and give a sound and complete axiomatic system for these counterfactuals.

The deterministic semantics of causal counterfactuals is provided by Halpern [14] in his causal reasoning logic. As conditionals, causal counterfactuals could have a probabilistic interpretation in chancy scenarios. Leitgeb [26] has proposed probabilistic semantics for counterfactuals, that is, a counterfactual is true in a world if and only if the corresponding world-relative conditional probability is sufficiently high. Similar proposals can be seen in [24], [4, chap. 16] and [5, pp. 213-5]. In the causal modelling framework, the truth condition of a causal counterfactual could be its corresponding probabilistic quantity being sufficiently high. In Pearl's probabilistic causal models, a *counterfactual probability* of the form $P(\varphi_{\mathbf{X}=\mathbf{x}} \mid e)$ is a conditional probability of a causal counterfactual $\mathbf{X} = \mathbf{x} \Box\!\!\rightarrow \varphi$. For instance, the quantity $P(FF = 0_{MD=0} \mid MD = FF = 1)$ is the probability of a counterfactual $MD = 0 \Box\!\!\rightarrow FF = 0$ based on facts $MD = FF = 1$. As counterfactual probabilities take actual situations into account and could relate to an actual world, these quantities are appropriate to be the truthmaker for causal counterfactuals in probabilistic causal models. Hence, the truth condition of a causal counterfactual in a "probabilistic causal world" could be the corresponding counterfactual probability being sufficiently high.

The term "being sufficiently high" needs to be more precise by setting an appropriate threshold. Leitgeb presented two semantics, one interprets "being sufficiently high" as "equal to the maximal value 1," the other gives "sufficiently high" an accurate real-valued threshold $1 - \alpha$ ($0 \leq \alpha < \frac{1}{2}$). Accordingly, we could have "equal to 1" semantics with the corresponding counterfactual probabilities being equal to 1. I provide a sound and complete axiomatic system for this semantics. Likewise, we could define the approximate truth of a causal counterfactual to degree $1 - \alpha$ in terms of its counterfactual probability exceeding or being equal to $1 - \alpha$ ($0 \leq \alpha < \frac{1}{2}$). I compare the validity of some common conditional properties of causal counterfactuals between the two semantics.

One application of the probabilistic semantics of causal counterfactuals is to interpret counterfactuals in counterfactual accounts of actual causation probabilistically, so that we obtain probabilistic versions of those accounts. This could be seen as an alternative approach to *probabilistic causation* compared with the traditional straightforward *probability raising* approach. I choose a plausible and useful account of actual cause defined by causal counterfactuals, the modified Halpern-Pearl definition [15]. It becomes an indeterministic account once counterfactuals are interpreted probabilistically. Unlike general theories of probabilistic causation following the probability raising principle, the distinct feature of the indeterministic accounts we obtain is the requirement of significant probability raising instead of just slight probability raising in other theories. I treat the new indeterministic accounts as the semantics of "causal conditionals" formulated as $\mathbf{X} = \mathbf{x} \xrightarrow{\mathbf{C}} \varphi$ and explore the validity of some causal and conditional properties to reflect the features of those probabilistic accounts.

The paper will proceed as follows. The target language causal counterfactuals and the setup of Pearl's probabilistic causal models will be introduced in the next section. I present the above-threshold semantics ("equal to 1" and "not less than 1-α" semantics) in Sect. 3 and provide completeness proof for "equal to 1" semantics. In Sect. 4, I introduce and compare other approaches to probabilistic causal logics, including Ibeling and Icard's probabilistic logic of three-level causal hierarchy [23], Beckers' probabilistic generalization of nondeterministic causal models [3], and Barbero and Virtema's logic of probabilistic interventionist counterfactuals based on causal multiteam semantics. The probabilistic semantics of causal counterfactuals is adopted into counterfactual theories of actual causation, and the relevant logical results are discussed in Sect. 5. I conclude in the last section.

2 Causal Counterfactual and Probabilistic Causal Model

Causal counterfactuals are the language derived from causal models that can be extended into probabilistic settings. I will first introduce causal models, the language and the deterministic semantics of it [14], and then present Pearl's probabilistic causal models [29, p. 205].

2.1 Causal Model and Causal Counterfactual

A causal model or structural equation model is associated with a *signature* S, which includes a set $\mathbf{U}$ of exogenous variables, a set $\mathbf{V}$ of endogenous variables and an assignment R that assigns each variable $X \in \mathbf{U} \cup \mathbf{V}$ a non-empty domain $R(X)$. The values of exogenous variables are given outside the model, and the value of each endogenous variable is determined by other variables. A causal model M over a signature S is a pair $\langle S, F\rangle$ in which F is a set of structural equations specifying the dependencies among variables. Each endogenous variable X has a structural equation $X = f_X(\mathbf{U} \cup \mathbf{V} \setminus \{X\})(f_X : \times_{Y \in (\mathbf{U} \cup \mathbf{V} \setminus \{X\})} R(Y) \rightarrow R(X))$ that determines a unique value for X considering the values of other variables. Exogenous variables do not depend on other variables and thus have no equations. A value assignment $\mathbf{u}$[1] to all exogenous variables $\mathbf{U}$ which is an abbreviation of $U_1 = u_1, U_2 = u_2, ...(U_i \in \mathbf{U}$ and $u_i \in R(U_i))$ is called a *context*, and a pair $\langle M, \mathbf{u}\rangle$ consisting of a causal model and a context is called a *causal setting*. Given a context $\mathbf{u}$, the equations have a unique solution $\mathbf{v}(\mathbf{u})$, that is, a context can determine the values of all endogenous variables.

[1] In this paper, the bold capital letters (e.g. $\mathbf{X}$) represent a set or a vector of variables, and the bold lowercase letters (e.g. $\mathbf{x}$) represent a value configuration of corresponding variables. For instance, $\mathbf{X} = \{X_1, ..., X_n\}$ or $\mathbf{X} = (X_1, ..., X_n)$; $\mathbf{X} = \mathbf{x}$ is an abbreviation of $X_1 = x_1, ..., X_n = x_n$ where $x_i \in R(X_i)$.

For illustration, consider Halpern's "forest fire" example [16, pp. 10, 14]: a forest fire can be triggered by lightning or a lit match dropped by an arsonist. A causal model M^{FF} for this case could include five binary variables, two exogenous and three endogenous. The exogenous variable U_1 represents whether the external conditions (e.g., humidity and temperature) are such that the lightning strikes, and U_2 represents whether the arsonist drops the match (psychological conditions). The endogenous variable L represents whether lightning strikes, MD represents whether a lit match is dropped, and FF represents whether a forest fire starts. For example, L takes value 1 if lightning strikes, 0 otherwise. The structural equations are $L = U_1$, $MD = U_2$ and $FF = L \vee MD$. Once the values of U_1 and U_2 are given, the values of endogenous variables are determined via equations.

An intervention on endogenous variables is understood as "equation replacement." Specifically, for a set of endogenous variables $\mathbf{X}$ and its value configuration $\mathbf{x} \in R(\mathbf{X})$ (or $\mathbf{x} \in \times_{X \in \mathbf{X}} R(X)$) in a causal model M, an intervention that forces $\mathbf{X}$ to attain $\mathbf{x}$ ($do(\mathbf{X} = \mathbf{x})$) is to replace the equation of any $X (\in \mathbf{X})$ with a constant function $X = x$ where x is the assignment to X in $\boldsymbol{x}$, the equations for variables not in $\mathbf{X}$ remain unchanged. This modification for M generates a new model $M_{\mathbf{X}=\boldsymbol{x}} = \langle S, F_{\mathbf{X}=\boldsymbol{x}} \rangle$ where $F_{\mathbf{X}=\boldsymbol{x}}$ is the set of equations after the replacement. Generally, intervention is defined on endogenous variables. For technical reasons, here we extend intervention to cover exogenous variables. An intervention on exogenous variables is just to reset their values in a context. In particular, an intervention $do(\mathbf{U_s} = \mathbf{u_s})$ ($\mathbf{U_s} \subseteq \mathbf{U}$, $\mathbf{u_s} \in R(\mathbf{U_s})$) in a context $\mathbf{u}$ is to substitute the value of each $U \in \mathbf{U_s}$ in $\mathbf{u_s}$ for its original value in $\mathbf{u}$. Such intervention forces a model to consider some specific background conditions. After any intervention $do(\mathbf{X} = \mathbf{x})$ ($\mathbf{X} \subseteq \mathbf{U} \cup \mathbf{V}$) in a causal setting $\langle M, \mathbf{u} \rangle$, we could use the notation $\langle M, \mathbf{u} \rangle_{\mathbf{X}=\mathbf{x}}$ to represent the modified model $M_{\mathbf{X_1}=\mathbf{x_1}}$ and context $\mathbf{u}_{\mathbf{X_2}=\mathbf{x_2}}$ where $\mathbf{X_1} \subseteq \mathbf{V}$, $\mathbf{X_2} \subseteq \mathbf{U}$, $\mathbf{X_1} \cup \mathbf{X_2} = \mathbf{X}$ and $\mathbf{x_1}$, $\mathbf{x_2}$ are the corresponding value configurations of $\mathbf{X_1}$, $\mathbf{X_2}$ in $\mathbf{x}$.

In forest fire example, we could intervene to drop a lit match $do(MD = 1)$ by replacing the equation of $MD = U_2$ with $MD = 1$ even in the context $U_1 = 1, U_2 = 0$, and then calculate the values of L and FF by their equations in that original context. The intervention $do(MD = 1)$ creates a counterfactual scenario so that a counterfactual conditional like $MD = 1 \;\Box\!\!\rightarrow FF = 1$ could be evaluated in $\langle M^{FF}_{MD=1}, (1, 0) \rangle$. If we intervene to set $U_1 = 0$ and $MD = 1$ in $\langle M^{FF}, (1,0) \rangle$, the model after the modification is simply $\langle M^{FF}_{MD=1}, (0, 0) \rangle$ with the value of U_1 in the original context replaced.

Accordingly, the formal language of causal counterfactuals I consider in this paper allows sentences with antecedents expressed by exogenous variables. Given a signature $S = \langle \mathbf{U}, \mathbf{V}, R \rangle$, a primitive formula is an endogenous variable taking its valuc (e.g. $Y = y, Y \in \mathbf{V}, y \in R(Y)$); a causal counterfactual is of the form $\mathbf{X} = \mathbf{x} \;\Box\!\!\rightarrow \varphi'$ where $\mathbf{X} = \mathbf{x}$ is an abbreviation of $X_1 = x_1 \wedge ... \wedge X_m = x_m$ ($X_1, ..., X_m$ are distinct variables, $X_i \in \mathbf{U} \cup \mathbf{V}, x_i \in R(X_i)$), and φ' is a

Boolean combination of primitive formulas.[2] It expresses the statement "if a set of variables $\mathbf{X}$ were set to its value configuration $\mathbf{x}$, then φ' would hold." The special case when $\mathbf{X}$ is empty is just φ'. The language contains Boolean combinations of causal counterfactuals, denoted as $\mathcal{L}(S)$.

Halpern [14] provided deterministic semantics for causal counterfactuals with respect to three classes of causal models. Here we focus on the models with the "unique-solution" property, that is, the equations in such a model have a unique solution under arbitrary interventions given any context.[3] We call such models *Pearlian* causal models. The satisfaction relation $\Vdash$ between formulas in $\mathcal{L}(S)$ and Pearlian models is defined as follows:

- $\langle M, \mathbf{u}\rangle \Vdash X = x$ iff X takes value x in the solution to M relative to $\mathbf{u}$;
- $\langle M, \mathbf{u}\rangle \Vdash \mathbf{X} = \mathbf{x} \,\Box\!\!\rightarrow \varphi'$ iff $\langle M, \mathbf{u}\rangle_{\mathbf{X}=\mathbf{x}} \Vdash \varphi'$;
- $\langle M, \mathbf{u}\rangle \Vdash \neg\varphi$ iff $\langle M, \mathbf{u}\rangle \nVdash \varphi$;
- $\langle M, \mathbf{u}\rangle \Vdash \varphi \wedge \psi$ iff $\langle M, \mathbf{u}\rangle \Vdash \varphi$ and $\langle M, \mathbf{u}\rangle \Vdash \psi$.

Note that the modified model in $\langle M, \mathbf{u}\rangle_{\mathbf{X}=\mathbf{x}}$ is still Pearlian according to the definition of Pearlian models.

2.2 Probabilistic Causal Model

According to Pearl's [29, p. 205] definition, a *probabilistic causal model* is a pair $\langle M, P\rangle$ where $M = \langle \mathbf{U}, \mathbf{V}, R, F\rangle$ is a Pearlian causal model and P is a probability distribution defined over the domain of $\mathbf{U}$. The probability of a causal counterfactual $P(\varphi'_{\mathbf{X}=\mathbf{x}})$ can be calculated by summing up the probabilities of the contexts where $\mathbf{X} = \mathbf{x} \,\Box\!\!\rightarrow \varphi'$ is true:

$$P(\varphi'_{\mathbf{X}=\mathbf{x}}) = \sum_{\{\mathbf{u}\in R(\mathbf{U}):\ \langle M, u\rangle \Vdash \mathbf{X}=\mathbf{x}\Box\!\!\rightarrow\varphi'\}} P(\mathbf{u}).$$

If we further condition on some observed facts, we obtain a conditional probability of a causal counterfactual, $P(\varphi'_{\mathbf{X}=\mathbf{x}} \mid \mathbf{Y} = \mathbf{y})$ $(\mathbf{Y} \subseteq \mathbf{V}, \mathbf{y} \in R(\mathbf{Y}))$, which is calculated as:

$$P(\varphi'_{\mathbf{X}=\mathbf{x}} \mid \mathbf{Y} = \mathbf{y}) = \sum_{\{\mathbf{u}\in R(\mathbf{U}):\ \langle M, \mathbf{u}\rangle \Vdash \mathbf{X}=\mathbf{x}\Box\!\!\rightarrow\varphi'\}} P(\mathbf{u} \mid \mathbf{Y} = \mathbf{y}).$$

That is, the probability of $\mathbf{X} = \mathbf{x} \,\Box\!\!\rightarrow \varphi'$ given $\mathbf{Y} = \mathbf{y}$ is the sum of the updated probabilities of contexts where the counterfactual is true. We call such quantities *counterfactual probabilities*. For instance, $P(Y = y'_{X=x'} \mid X = x, Y = y)$ is the probability that Y would not take value y had X not been set to x, given that $X = x$ and $Y = y$ have in fact occurred. Counterfactual probabilities

[2] Other expressions include $[\mathbf{X} \leftarrow \mathbf{x}]\varphi$ [16] and $\mathbf{Y}_{\mathbf{x}} = \mathbf{y}$ or its abbreviation $\mathbf{y}_{\mathbf{x}}$ which is often used in the statistics literature on causal inference.

[3] The unique-solution property still holds for such a model if interventions involve exogenous variables.

are capable of evaluating the possibility of counter-to-fact situations by taking actually occurring facts into account.

We could turn the forest fire example into its probabilistic version. Now it is not necessarily the case that a forest fire starts once the arsonist drops a lit match or lightning strikes. Suppose that: If lightning does not occur but an arsonist drops a lit match, there is a 0.5 chance that a forest fire starts; If lightning strikes and there is no arsonist dropping a lit match, the chance of a forest fire is 0.4; Given that lightning does not occur and there is no arsonist dropping a lit match, the chance of a forest fire is 0.1 (since other factors like shifting cultivation can cause a forest fire); The chance of the occurrence of lightning is 0.3; The chance of dropping a lit match by an arsonist is 0.2. In the actual situation, the arsonist dropped a lit match and lightning struck, a forest fire started.

To construct a probabilistic causal model $\langle M^{FF'}, P\rangle$ for this example, first we modify M^{FF} to incorporate more exogenous variables to represent chancy dependencies between endogenous variables. The equation of FF could be written as $FF = (L \wedge U_3) \vee (MD \wedge U_4) \vee U_5$ where U_3 represents the conditions under which once lightning strikes, a forest fire starts (e.g. precipitation and the amount of the wood); U_4 represents the conditions under which once a lit match is dropped, a forest fire starts (e.g. the dryness of the wood and the amount of oxygen); U_5 represents other factors that can trigger a forest fire (e.g. a volcanic eruption and shifting cultivation). According to the description of the example and the equations of FF, $L(L = U_1)$ and $MD(MD = U_2)$ in the new causal model $M^{FF'}$, the relations among the endogenous variables become indeterministic and can simply be represented by the probabilities of the exogenous variables:

- $P(U_1 = 1) = P(L = 1) = 0.3$;
- $P(U_2 = 1) = P(MD = 1) = 0.2$;
- $P(U_3 = 1 \vee U_5 = 1) = P(FF = 1 \mid L = 1, MD = 0) = 0.4$;
- $P(U_4 = 1 \vee U_5 = 1) = P(FF = 1 \mid L = 0, MD = 1) = 0.5$;
- $P(U_5 = 1) = P(FF = 1 \mid L = 0, MD = 0) = 0.1$.

Assume that the exogenous variables are mutually independent, the probability distribution $P(u_1, ..., u_5) = \prod_{i \leqslant 5} P(u_i)(u_i \in R(U_i))$. We can calculate $P(FF = 1) = \sum_{\{\mathbf{u} \mid \langle M^{FF'}, \mathbf{u}\rangle \Vdash FF=1\}} P(\mathbf{u}) = 0.262$ and $P(FF = 1_{MD=0}) = \sum_{\{\mathbf{u} \mid \langle M^{FF'}, \mathbf{u}\rangle \Vdash MD=0 \Box\!\!\rightarrow FF=1\}} P(\mathbf{u}) = 0.19$. As the actual facts are $L = MD = FF = 1$, the probability of the counterfactual "if the arsonist had not dropped the lit match, the forest fire would still have happened" given these facts can be evaluated as $P(FF = 1_{MD=0} \mid L = MD = FF = 1) = \sum_{\{\mathbf{u} \mid \langle M^{FF'}, \mathbf{u}\rangle \Vdash MD=0 \Box\!\!\rightarrow FF=1\}} P(\mathbf{u} \mid L = MD = FF = 1) = 0.6$. Specifically, given $L = MD = FF = 1$, we obtain the facts that $U_1 = 1, U_2 = 1, U_3 = 1 \vee U_4 = 1 \vee U_5 = 1$ by abduction based on the equations. Then we update the distribution over the contexts that satisfy the above facts and verify whether $MD = 0 \Box\!\!\rightarrow FF = 1$ is true in those contexts.

To determine the truth values of counterfactuals, we could define a *probabilistic causal world* $\langle M, P, \mathbf{u}\rangle$ in which $\langle M, P\rangle$ is a probabilistic causal model and

$\mathbf{u}$ is a possible context that could derive an actual setting of all the endogenous variables $\mathbf{v}(\mathbf{u})$.[4] It seems that $\langle M, P, \mathbf{u}\rangle$ is a deterministic world. However, $\mathbf{u}$ is one of the contexts which are consistent with some $\mathbf{v}$ that represents observed facts. A solution $\mathbf{v}$ can be derived by several different contexts in a Pearlian causal model. The actual context is unknown, which brings about uncertainty into the world.

The reason why I use exogenous variables to bear the uncertainty is the conceptual distinction between exogenous and endogenous variables. Exogenous variables are usually understood as representing unobservable factors [29, pp. 69, 203, 274]. In general, like in causal examples, all the relevant factors are represented by endogenous variables that are taken as observable. Therefore, an actual value setting of the endogenous variables encompasses all of the relevant facts of actuality, and can be regarded as known and observed information of a world. The uncertainty in a probabilistic world then is produced by unobserved or background variables $\mathbf{U}$ [29, pp. 26, 220]. The actual solution $\mathbf{v}$ is also suitable to be presented in a probabilistic world, for simplicity, I choose contexts $\mathbf{u}$ as the element in the world.

3 Probabilistic Causal Logic

The probabilistic logic for causal counterfactuals I propose is relative to a signature $S = \langle \mathbf{U}, \mathbf{V}, R\rangle$ in which $\mathbf{U}$, $\mathbf{V}$ and the domain $R(X)$ of each variable $X \in \mathbf{U} \cup \mathbf{V}$ are finite. I name the logic "Probabilistic Causal Logic", hereafter called **PCL(S)**. The language of **PCL(S)** is defined in Sect. 2.1, denoted as $\mathcal{L}(S)$. Probabilistic causal models I consider are those that assign a positive probability to each context (i.e. for any $\boldsymbol{u}$, $P(\boldsymbol{u}) > 0$).[5] Let $\mathcal{C}(S)$ be the class of all such models.[6]

In probabilistic causal models, counterfactual probabilities take observations into consideration, and are appropriate quantities to specify the truth condition of a causal counterfactual. For a probabilistic world, the observed facts are the

[4] $\mathbf{v}$ can also be a part of a probabilistic world $\langle M, P, \mathbf{v}\rangle$ with the restriction that $P(\mathbf{v}) > 0$ for defining the semantics of counterfactuals.

[5] This restriction ensures that the counterfactual probabilities we use in the semantics are well-defined and is to guarantee the validity of axiom TRANSFORMATION. It is stronger than the requirement "regular". A probability function is regular if it assigns probability 1 only to necessary propositions, and 0 only to impossible propositions [12, p. 280].

[6] Except the property of unique solution, we can also consider probabilistic causal models equipped with recursive (or acyclic) causal models. Recursive causal models are Pearlian, so the class of probabilistic causal models with recursive models is a subclass of probabilistic causal models defined in Sect. 2.2. The axiomatic system with respect to this subclass can be obtained by adding a characteristic axiom for recursiveness (the same as axiom "recursiveness" in the deterministic system for causal counterfactuals relative to recursive models provided by Halpern [14]) to system **SPCL(S)** which will be presented below. The detailed introduction of recursive causal models and its deterministic axiom system can be found in [14].

actual setting of all the endogenous variables, which ought to be conditioned upon to provide as many relevant information as possible. I refer to Leitgeb [26]'s probabilistic semantics of counterfactuals and present "equal to 1" semantics of **PCL(S)** first as follows.

Definition 1 ($\vDash$)**.** *The satisfaction relation* $\vDash$ *between a probabilistic causal world* $\langle M, P, \boldsymbol{u}\rangle$ $(\langle M, P\rangle \in \mathcal{C}(S))$ *and a formula* φ $(\in \mathcal{L}(S))$ *is as below:*[7]

$$\begin{array}{|lcl|}\hline \langle M, P, \boldsymbol{u}\rangle \vDash \boldsymbol{X} = \boldsymbol{x} \mathrel{\Box\!\!\rightarrow} \varphi' & \Longleftrightarrow & P(\varphi'_{\boldsymbol{X}=\boldsymbol{x}} \mid \boldsymbol{v}(\boldsymbol{u})) = 1 \\ \langle M, P, \boldsymbol{u}\rangle \vDash \neg\varphi & \Longleftrightarrow & \langle M, P, \boldsymbol{u}\rangle \nvDash \varphi \\ \langle M, P, \boldsymbol{u}\rangle \vDash (\varphi \wedge \psi) & \Longleftrightarrow & \langle M, P, \boldsymbol{u}\rangle \vDash \varphi \text{ and } \langle M, P, \boldsymbol{u}\rangle \vDash \psi \\ \hline \end{array}$$

A formula φ *is valid on* $\langle M, P\rangle$ $(\langle M, P\rangle \vDash \varphi)$ *if and only if for all* $\boldsymbol{u}$, $\langle M, P, \boldsymbol{u}\rangle \vDash \varphi$.

Based on the above semantics, the truth of a formula φ' depends on whether it is consistent with the actual setting of all endogenous variables $\boldsymbol{v}(\boldsymbol{u})$. The restriction on probabilistic causal models (each context has a positive probability)

[7] According to Leitgeb [26, pp. 60-3], counterfactuals are typically sentences which describe the world and are true or false, the probability measures that interpret them should be objective nonepistemic ones. Nonetheless, he doesn't reject subjective or epistemic interpretations of single case chances in his semantics. Actually, the probabilities in probabilistic causal models are introduced because of our ignorance of the details that could specify an actual world, so they ought to be regarded as subjective or epistemic [29, pp. 26, 310]. It appears that probabilistic causal models are not suitable setups to provide semantics for counterfactuals. However, counterfactual probabilities are the corresponding probability quantities for causal counterfactuals, and probabilistic causal models have the concise evaluation methodology for counterfactual probabilities. It seems that probabilistic causal models are the only choice for probabilistic semantics of causal counterfactuals. In order to mitigate the worry that counterfactuals are commonly thought of as ontic whereas probabilities in probabilistic causal models characterize mental state of uncertainty, as causal models are abstract mathematical constructs, we can regard exogenous variables as representational devices to bear randomness (see the example "probabilistic forest fire" in Sect. 2.2), or in other words, they can be viewed as fictional errors rather than substantial factors in some sense so that probabilistic causal models can be seen as aiming to modelling nondeterministic dependencies (represented by endogenous variables) essentially just as causal Bayesian networks do by ignoring exogenous variables. Such view about exogenous variables is similar to the viewpoint of "pseudo-hidden variable" named by Steel [31], according to which, exogenous variables do not denote causes or factors but represent fundamentally nondeterministic causal processes. This perspective makes it possible that the probabilities in probabilistic causal models are interpreted as objective chances that characterize stochastic laws of nature or causal processes like probabilities in causal Bayesian networks. Anyway, we can also bite the bullet, and accept an epistemic reading of causal counterfactuals so that they are understood as special conditionals which are between being ontic and being epistemic. I will not discuss more about the interpretation of probability in this paper and will focus on the technical results of the probabilistic logic.

makes "equal to 1" semantics of causal counterfactuals less probabilistic. The condition $P(\varphi'_{\boldsymbol{X}=\boldsymbol{x}} \mid \boldsymbol{v}(\boldsymbol{u})) = 1$ actually requires that all the contexts consistent with $\boldsymbol{v}$ satisfy $\boldsymbol{X} = \boldsymbol{x} \Box\!\!\rightarrow \varphi'$ given the above restriction on models. In other words, the first clause can change into "$\langle M, P, \boldsymbol{u}\rangle \vDash \boldsymbol{X} = \boldsymbol{x} \Box\!\!\rightarrow \varphi' \iff \forall \boldsymbol{u}'\ s.t.\ \langle M, \boldsymbol{u}'\rangle \Vdash \boldsymbol{V} = \boldsymbol{v}$ (where $\langle M, \boldsymbol{u}\rangle \Vdash \boldsymbol{V} = \boldsymbol{v}$), $\langle M, \boldsymbol{u}'\rangle \Vdash \boldsymbol{X} = \boldsymbol{x} \Box\!\!\rightarrow \varphi'$," which implies that probability doesn't work in this semantics. Later in this section, I will introduce "$1 - \alpha$" semantics which makes use of probability substantially. For now, we proceed with the axiomatization of **PCL(S)** relative to "equal to 1" semantics.

At first, some notations we use in the following need to be explained. A sequence of variables $\boldsymbol{X_Y}$ or $\boldsymbol{X}_{\{Y_1,\dots,Y_k\}}$ means the sequence of variables obtained by deleting $\boldsymbol{Y}$ or $\{Y_1, \dots, Y_k\}$ from $\boldsymbol{X}$. Likewise, a sequence of values $\boldsymbol{x_Y}$ or $\boldsymbol{x}_{\{Y_1,\dots,Y_k\}}$ is the sequence of values obtained by deleting the values of $\boldsymbol{Y}$ or $\{Y_1, \dots, Y_k\}$ in $\boldsymbol{x}$. Thus $\boldsymbol{X_Y} = \boldsymbol{x_Y}$ or $\boldsymbol{X}_{\{Y_1,\dots,Y_k\}} = \boldsymbol{x}_{\{Y_1,\dots,Y_k\}}$ is the conjunction of primitive formulas derived from removing $\boldsymbol{Y} = \boldsymbol{y}$ ($\boldsymbol{y}$ is the values corresponding to $\boldsymbol{Y}$ in $\boldsymbol{x}$) from $\boldsymbol{X} = \boldsymbol{x}$. If only one variable Y needs to be removed, we simply write those notations as $\boldsymbol{X}_Y$, $\boldsymbol{x}_Y$ and $\boldsymbol{X}_Y = \boldsymbol{x}_Y$.

System **SPCL(S)**

Axiom Schemas

PROP	Classical propositional axioms
DEFINITENESS	$\bigvee_{x \in R(X)} X = x$
FUNCTIONALITY	$\bigvee_{x \in R(X)} (\boldsymbol{U} = \boldsymbol{u} \wedge \boldsymbol{Y} = \boldsymbol{y}) \Box\!\!\rightarrow X = x$
EQUALITY	$\mathbf{Y} = \mathbf{y} \Box\!\!\rightarrow X = x \Rightarrow \neg \mathbf{Y} = \mathbf{y} \Box\!\!\rightarrow X = x'$ where $x \neq x'$
EFFECTIVENESS	$(\mathbf{Y} = \mathbf{y} \wedge X = x) \Box\!\!\rightarrow X = x$
CONJUNCTION	$\mathbf{Y} = \mathbf{y} \Box\!\!\rightarrow (\varphi' \wedge \psi') \Leftrightarrow \mathbf{Y} = \mathbf{y} \Box\!\!\rightarrow \varphi' \wedge \mathbf{Y} = \mathbf{y} \Box\!\!\rightarrow \psi'$
DISJUNCTION	$(\mathbf{U} = \mathbf{u} \wedge \mathbf{Y} = \mathbf{y}) \Box\!\!\rightarrow (\varphi' \vee \psi') \Leftrightarrow$ $(\mathbf{U} = \mathbf{u} \wedge \mathbf{Y} = \mathbf{y}) \Box\!\!\rightarrow \varphi' \vee (\mathbf{U} = \mathbf{u} \wedge \mathbf{Y} = \mathbf{y}) \Box\!\!\rightarrow \psi'$
COMPOSITION	$\mathbf{Y} = \mathbf{y} \Box\!\!\rightarrow (X = x \wedge \mathbf{Z} = \mathbf{z}) \Rightarrow (\mathbf{Y} = \mathbf{y} \wedge X = x) \Box\!\!\rightarrow \mathbf{Z} = \mathbf{z}$
CAUSAL WORLD	$\mathbf{V} = \mathbf{v} \Rightarrow \bigvee_{\mathbf{u} \in R(\mathbf{U})} \mathbf{U} = \mathbf{u} \Box\!\!\rightarrow \mathbf{V} = \mathbf{v}$
TRANSFORMATION	$\mathbf{V} = \mathbf{v} \Rightarrow \mathbf{Y} = \mathbf{y} \Box\!\!\rightarrow \varphi' \Leftrightarrow$ $\bigwedge_{\mathbf{u} \in R(\mathbf{U})} (\mathbf{U} = \mathbf{u} \Box\!\!\rightarrow \mathbf{V} = \mathbf{v} \Rightarrow (\mathbf{U} = \mathbf{u} \wedge \mathbf{Y} = \mathbf{y}) \Box\!\!\rightarrow \varphi')$
SOLUTION	$(\mathbf{U} = \mathbf{u} \wedge \mathbf{Y} = \mathbf{y} \wedge \mathbf{V}_{\mathbf{Y} \cup \{X_1\}} = \mathbf{v}_{\mathbf{Y} \cup \{X_1\}}) \Box\!\!\rightarrow X_1 = x_1 \wedge \dots \wedge$ $(\mathbf{U} = \mathbf{u} \wedge \mathbf{Y} = \mathbf{y} \wedge \mathbf{V}_{\mathbf{Y} \cup \{X_m\}} = \mathbf{v}_{\mathbf{Y} \cup \{X_m\}}) \Box\!\!\rightarrow X_m = x_m \Rightarrow$ $(\mathbf{U} = \mathbf{u} \wedge \mathbf{Y} = \mathbf{y}) \Box\!\!\rightarrow \mathbf{V} = \mathbf{v}$ where $\mathbf{Y} \cup \{X_1, \dots, X_m\} = \mathbf{V}$, $\mathbf{Y} \cap \{X_1, \dots, X_m\} = \emptyset$, $\mathbf{y}$ and x_i are the value settings of $\mathbf{Y}$ and X_i in $\mathbf{v}$ respectively

Rules

MP	From $\vdash \varphi \Rightarrow \psi$ and $\vdash \varphi$, infer $\vdash \psi$
RE	From $\vdash \varphi' \Leftrightarrow \psi'$, infer $\vdash \mathbf{Y} = \mathbf{y} \Box\!\!\rightarrow \varphi' \Leftrightarrow \mathbf{Y} = \mathbf{y} \Box\!\!\rightarrow \psi'$

DEFINITENESS says that each endogenous variable in a probabilistic causal world will take on a certain value. FUNCTIONALITY tells us that in a deterministic context $\boldsymbol{u}$, each endogenous variable has a certain value after any intervention on endogenous variables. DEFINITENESS and FUNCTIONALITY are based on the assumption that for any $X \in \mathbf{V}$, $R(X)$ is finite. FUNCTIONALITY involves $\mathbf{U}$ in the formula which is guaranteed by the assumption that $\mathbf{U}$ is finite, same with DISJUNCTION, CAUSAL WORLD, TRANS-

FORMATION and SOLUTION. EQUALITY guarantees that the value of each endogenous variable after any intervention is unique if that variable has a value under that intervention. When $\boldsymbol{Y}$ in EQUALITY is an empty set, that gives us $X = x \Rightarrow \neg X = x'$, then with DEFINITENESS, it implies that each endogenous variable takes a unique value in a probabilistic causal world. If we have $\boldsymbol{U} = \boldsymbol{u} \,\Box\!\!\rightarrow X = x_1 \Rightarrow \neg \boldsymbol{U} = \boldsymbol{u} \,\Box\!\!\rightarrow X = x_2$ according to EQUALITY, together with an instance of FUNCTIONALITY $\bigvee_{x \in R(X)} \boldsymbol{U} = \boldsymbol{u} \,\Box\!\!\rightarrow X = x$, actually we switch from a probabilistic world to a deterministic world $\langle M, \boldsymbol{u} \rangle$ and obtain the property of causal models that each endogenous variable has a unique value in a context. CONJUNCTION makes it possible that any causal counterfactual with a conjunction as its consequent can be split into the conjunction of several counterfactuals with the conjuncts of the original consequent serving as their new consequents respectively. DISJUNCTION also helps us to separate a disjunctive statement after some certain intervention, but it works only in deterministic worlds which can be realized by intervening on $\mathbf{U}$. CAUSAL WORLD says that a solution in a probabilistic world has a context that could derive that solution. TRANSFORMATION translates the truth condition of $\boldsymbol{Y} = \boldsymbol{y} \,\Box\!\!\rightarrow \varphi'$ on a probabilistic causal world. The fact that the range of each exogenous variable is finite ensures that CAUSAL WORLD and TRANSFORMATION are formulas in the sense that infinite disjunction and conjunction are avoided. SOLUTION reflects a characteristic of a solution $\boldsymbol{v}$ in a context $\boldsymbol{u}$ under any intervention on endogenous variables $\mathbf{Y} = \mathbf{y}$. That is, equations are consistent by assigning the corresponding values of $\boldsymbol{v}$ to the variables in equations of $\mathbf{V_Y}$, or in other words, $\boldsymbol{v}$ solves the equations of $\mathbf{V_Y}$. This axiom expresses that once there is a consistent result on equations under any intervention given a context, the values of endogenous variables assigned to the equations constitute a solution under that intervention in the context. CAUSAL WORLD, TRANSFORMATION and SOLUTION involve $\boldsymbol{V} = \boldsymbol{v}$ which requires that $\mathbf{V}$ is finite.

Although probability does not essentially work in the semantics due to the restriction on models (i.e. the probability of each context is positive), the implicit role played by probability is to consider many other contexts except the "actual" one. The fact that TRANSFORMATION is invalid for Halpern's deterministic causal logic [14] also reflects that our semantics involves other contexts instead of just a single context due to the existence of probability distribution.[8] At that point, **SPCL(S)** cannot be seen as a variant of Halpern's deterministic causal logic for non-probabilistic causal models.[9]

Theorem 1. ***SPCL(S)*** *is sound over* $\mathcal{C}(S)$.

Proof. The proofs are straightforward.

To prove the completeness of this system, we need to derive some theorems from the axioms and rules:

[8] For Halpern's deterministic semantics, $\mathbf{V} = \mathbf{v} \Rightarrow \mathbf{Y} = \mathbf{y} \,\Box\!\!\rightarrow \varphi' \Rightarrow \bigwedge_{\mathbf{u} \in R(\mathbf{U})} (\mathbf{U} = \mathbf{u} \,\Box\!\!\rightarrow \mathbf{V} = \mathbf{v} \Rightarrow (\mathbf{U} = \mathbf{u} \wedge \mathbf{Y} = \mathbf{y}) \,\Box\!\!\rightarrow \varphi')$ is not valid.

[9] Thank one reviewer for raising this issue.

Lemma 1. *For the following two theorems of* $\boldsymbol{SPCL(S)}$*, let* $\mathbf{Y} \cup \{X_1, ..., X_m\} = \mathbf{V}$*,* $\mathbf{Y} \cap \{X_1, ..., X_m\} = \emptyset$*,* $\mathbf{y}$ *and* x_i *be the value settings of* $\mathbf{Y}$ *and* X_i *in* $\mathbf{v}$ *respectively.*

1. *SOLUTION':* $\vdash (\mathbf{U} = \mathbf{u} \wedge \mathbf{Y} = \mathbf{y}) \Box\!\!\rightarrow \mathbf{V} = \mathbf{v} \Rightarrow (\mathbf{U} = \mathbf{u} \wedge \mathbf{Y} = \mathbf{y} \wedge \mathbf{V}_{\mathbf{Y} \cup \{X_1\}} = \mathbf{v}_{\mathbf{Y} \cup \{X_1\}}) \Box\!\!\rightarrow X_1 = x_1 \wedge ... \wedge (\mathbf{U} = \mathbf{u} \wedge \mathbf{Y} = \mathbf{y} \wedge \mathbf{V}_{\mathbf{Y} \cup \{X_m\}} = \mathbf{v}_{\mathbf{Y} \cup \{X_m\}}) \Box\!\!\rightarrow X_m = x_m$.
2. *POSSIBILITY:* $\vdash \boldsymbol{Y} = \boldsymbol{y} \Box\!\!\rightarrow \boldsymbol{V} = \boldsymbol{v} \Rightarrow \bigvee_{\boldsymbol{u} \in R(\boldsymbol{U})}((\boldsymbol{U} = \boldsymbol{u} \wedge \boldsymbol{Y} = \boldsymbol{y} \wedge \boldsymbol{V}_{\boldsymbol{Y} \cup \{X_1\}} = \boldsymbol{v}_{\boldsymbol{Y} \cup \{X_1\}}) \Box\!\!\rightarrow X_1 = x_1 \wedge ... \wedge (\boldsymbol{U} = \boldsymbol{u} \wedge \boldsymbol{Y} = \boldsymbol{y} \wedge \boldsymbol{V}_{\boldsymbol{Y} \cup \{X_m\}} = \boldsymbol{v}_{\boldsymbol{Y} \cup \{X_m\}}) \Box\!\!\rightarrow X_m = x_m)$.

Proof. See the appendix.

If we combine SOLUTION and SOLUTION', we have theorem SOLUTION* $(\mathbf{U} = \mathbf{u} \wedge \mathbf{Y} = \mathbf{y}) \Box\!\!\rightarrow \mathbf{V} = \mathbf{v} \Leftrightarrow (\mathbf{U} = \mathbf{u} \wedge \mathbf{Y} = \mathbf{y} \wedge \mathbf{V}_{\mathbf{Y} \cup \{X_1\}} = \mathbf{v}_{\mathbf{Y} \cup \{X_1\}}) \Box\!\!\rightarrow X_1 = x_1 \wedge ... \wedge (\mathbf{U} = \mathbf{u} \wedge \mathbf{Y} = \mathbf{y} \wedge \mathbf{V}_{\mathbf{Y} \cup \{X_m\}} = \mathbf{v}_{\mathbf{Y} \cup \{X_m\}}) \Box\!\!\rightarrow X_m = x_m$, which interprets a solution as a value setting of endogenous variables that makes all equations consistent after inputting it into the variables in equations.

We prove the completeness of $\mathbf{SPCL(S)}$ following the canonical model approach. That is, we show that any $\mathbf{SPCL(S)}$-consistent formula φ can be satisfiable. Consider a maximal consistent set Γ that includes φ. Because of classical propositional axioms and MP in the system, a maximal $\mathbf{SPCL(S)}$-consistent set Γ' has the following properties: for any $\mathcal{L}(S)$ formula ψ and ψ', (1) $\psi \notin \Gamma'$ if and only if $\neg\psi \in \Gamma'$; (2) $\psi \in \Gamma'$, $\psi' \in \Gamma'$ if and only if $\psi \wedge \psi' \in \Gamma'$; (3) $\psi \in \Gamma'$ or $\psi' \in \Gamma'$ if and only if $\psi \vee \psi' \in \Gamma'$.

We need to construct a canonical model $\langle M^c, P^c, \boldsymbol{u}^c \rangle$ such that for any ψ, $\psi \in \Gamma$ if and only if $\langle M^c, P^c, \boldsymbol{u}^c \rangle \vDash \psi$. Then we have $\langle M^c, P^c, \boldsymbol{u}^c \rangle \vDash \varphi$. To make the notation simpler, henceforth I sometimes omit variables in formulas. For instance, I write $\boldsymbol{y} \Box\!\!\rightarrow \boldsymbol{x}$ instead of $\boldsymbol{Y} = \boldsymbol{y} \Box\!\!\rightarrow \boldsymbol{X} = \boldsymbol{x}$.

Definition 2 (Canonical Model). *The canonical model* $\langle M^c, P^c, \boldsymbol{u}^c \rangle$ *is defined as follows:*

- $M^c = \langle S, F^c \rangle$ *where for each* $X \in \mathbf{V}$*,* f_X *is given as below:* *for any* $\boldsymbol{u} \in R(\boldsymbol{U})$ *and* $\boldsymbol{v}_X \in R(\boldsymbol{V}_X)$*,* $f_X(\boldsymbol{u}, \boldsymbol{v}_X) = x$ *if* $(\boldsymbol{u} \wedge \boldsymbol{v}_X) \Box\!\!\rightarrow X = x \in \Gamma$.
- P^c *is uniformly distributed over all the contexts* $\boldsymbol{u}$*, and* $\boldsymbol{u}^c$ *is any context that satisfies* $(\boldsymbol{u}^c \Box\!\!\rightarrow \boldsymbol{v}^c) \wedge \boldsymbol{v}^c \in \Gamma$.

Note that f_X is well-defined as $(\boldsymbol{u} \wedge \boldsymbol{v}_X) \Box\!\!\rightarrow X = x \in \Gamma$ is guaranteed by FUNCTIONALITY and EQUALITY. The existence of $\boldsymbol{u}^c$ depends on the condition $(\boldsymbol{u}^c \Box\!\!\rightarrow \boldsymbol{v}^c) \wedge \boldsymbol{v}^c \in \Gamma$. DEFINITENESS ensures that there is a solution $\boldsymbol{v}^c \in \Gamma$. Hereinafter we let $|\mathbf{V}| = n$. Due to POSSIBILITY, we have $\bigvee_{\boldsymbol{u} \in R(\boldsymbol{U})}((\boldsymbol{U} = \boldsymbol{u} \wedge \boldsymbol{V}_{X_1} = \boldsymbol{v}_{X_1}) \Box\!\!\rightarrow X_1 = x_1 \wedge ... \wedge (\boldsymbol{U} = \boldsymbol{u} \wedge \boldsymbol{V}_{X_n} = \boldsymbol{v}_{X_n}) \Box\!\!\rightarrow X_n = x_n) \in \Gamma$ $((\boldsymbol{v}_{X_i}, x_i) = \boldsymbol{v}^c)$, then we choose one context $\boldsymbol{u}^c$ s.t. $(\boldsymbol{U} = \boldsymbol{u}^c \wedge \boldsymbol{V}_{X_1} = \boldsymbol{v}_{X_1}) \Box\!\!\rightarrow X_1 = x_1 \wedge ... \wedge (\boldsymbol{U} = \boldsymbol{u}^c \wedge \boldsymbol{V}_{X_n} = \boldsymbol{v}_{X_n}) \Box\!\!\rightarrow X_n = x_n \in \Gamma$. According to SOLUTION*, we have $\boldsymbol{u}^c \Box\!\!\rightarrow \boldsymbol{v}^c \in \Gamma$, thus $(\boldsymbol{u}^c \Box\!\!\rightarrow \boldsymbol{v}^c) \wedge \boldsymbol{v}^c \in \Gamma$.

Next we prove that $\langle M^c, P^c, \boldsymbol{u}^c \rangle \in \mathcal{C}(S)$, which means M^c should be a Pearlian causal model and P^c is supposed to assign a positive probability to each context. As $\mathbf{U}$ and $R(U_i)$ for any $U_i \in \mathbf{U}$ are finite, suppose $|R(U_1)| \cdot \ldots \cdot |R(U_l)| = k$ $(l = |\mathbf{U}|)$, then for each context $\boldsymbol{u}$, $P^c(\boldsymbol{u}) = \frac{1}{k} > 0$. With this result, we only need to show that M^c is a Pearlian causal model so that $\langle M^c, P^c \rangle$ is a probabilistic causal model. It will follow that $\langle M^c, P^c \rangle \in \mathcal{C}(S)$.

Proposition 1. *M^c is a Pearlian causal model.*

Proof. The spirit of this proof is from [14, pp. 330-1]. See the appendix for detailed proof.

Theorem 2 (Truth Lemma). *For any $\psi \in \mathcal{L}(S)$, $\psi \in \Gamma$ iff $\langle M^c, P^c, \boldsymbol{u}^c \rangle \models \psi$.*

Proof. See the appendix.

Theorem 3. ***SPCL(S)** is complete with respect to $\mathcal{C}(S)$.*

Proof. Here we prove that every valid formula is a **SPCL(S)** theorem. That is, we show that for any $\varphi \in \mathcal{L}(S)$, if for all $\langle M, P \rangle \in \mathcal{C}(S)$, $\langle M, P \rangle \models \varphi$, then $\vdash \varphi$. We argue by contradiction. Suppose $\nvdash \varphi$, then $\neg\varphi$ is **SPCL(S)**-consistent. Consider a maximal consistent set Γ that includes $\neg\varphi$, according to Truth Lemma, $\neg\varphi$ can be satisfied. Thus φ is not valid, contradiction.

Now we turn to the other probabilistic causal logic denoted as $\mathbf{PCL(S)}^\alpha = \langle \mathcal{L}(S), \mathcal{C}(S), \models^\alpha \rangle$ with "1-α" semantics.

Definition 3 ($\models^\alpha$)**.** *The satisfaction relation $\models^\alpha$ $(0 \leq \alpha < \frac{1}{2})$ between a $\mathcal{L}(S)$ formula and a probabilistic causal world $\langle M, P, \boldsymbol{u} \rangle (\langle M, P \rangle \in \mathcal{C}(S))$ is defined as below:*

$$
\begin{array}{|lcl|}
\hline
\langle M, P, \boldsymbol{u} \rangle \models^\alpha \boldsymbol{X} = \boldsymbol{x} \mathrel{\Box\!\!\rightarrow} \varphi' & \Longleftrightarrow & P(\varphi'_{\boldsymbol{X}=\boldsymbol{x}} \mid \boldsymbol{v}(\boldsymbol{u})) \geq 1 - \alpha \\
\langle M, P, \boldsymbol{u} \rangle \models^\alpha \neg\varphi & \Longleftrightarrow & \langle M, P, \boldsymbol{u} \rangle \not\models^\alpha \varphi \\
\langle M, P, \boldsymbol{u} \rangle \models^\alpha (\varphi \wedge \psi) & \Longleftrightarrow & \langle M, P, \boldsymbol{u} \rangle \models^\alpha \varphi \text{ and } \langle M, P, \boldsymbol{u} \rangle \models^\alpha \psi \\
\hline
\end{array}
$$

A formula φ is valid on $\langle M, P \rangle$ ($\langle M, P \rangle \models^\alpha \varphi$) if and only if for all $\boldsymbol{u}$, $\langle M, P, \boldsymbol{u} \rangle \models^\alpha \varphi$.

If we refer to **SPCL(S)** and give an axiomatic system for this semantics, the axioms (including the rules) except CONJUNCTION and TRANSFORMATION are still valid. In terms of CONJUNCTION, $\boldsymbol{Y} = \boldsymbol{y} \mathrel{\Box\!\!\rightarrow} (\varphi' \wedge \psi') \Rightarrow \boldsymbol{Y} = \boldsymbol{y} \mathrel{\Box\!\!\rightarrow} \varphi' \wedge \boldsymbol{Y} = \boldsymbol{y} \mathrel{\Box\!\!\rightarrow} \psi'$ is valid, while the other direction is not. TRANSFORMATION is invalid, the translation of this new semantics for counterfactuals is as follows: the updated probability sum of contexts that are consistent with an "actual" $\mathbf{v}$ and make $\boldsymbol{Y} = \boldsymbol{y} \mathrel{\Box\!\!\rightarrow} \varphi'$ hold true is not less than $1 - \alpha$. Our language is not expressive enough to state the new semantics straightforwardly without probability.

The logical system with respect to Leitgeb's [26, pp. 54-5] "maximally high" semantics is complete, whereas the system for his "$1 - \alpha$" semantics [26, p. 58] seems unlikely to be complete based on the results from [28]. Likewise, the works on probabilistic above-threshold logics mentioned by Leitgeb (i.e. [19] and [20]) regard the relevant completeness problem as open questions.

The languages of [19,20,26] are counterfactuals and consequence relations respectively. For $\mathbf{PCL(S)}^\alpha$, it seems that the expressions with probability functions need to be added into the language to guarantee the completeness. Other than that, it is unclear whether there is another way to address this problem. Ibeling and Icard [23] have proposed axiomatizations of "probabilistic counterfactuals", i.e. probabilistic statements of causal counterfactuals (e.g. $P(\mathbf{X} = \mathbf{x} \Box\!\!\rightarrow \varphi) \geq P(\mathbf{Y} = \mathbf{y} \Box\!\!\rightarrow \psi))$ relative to probabilistic causal models. Our language can be extended by adding their probabilistic counterfactuals, so that we could straightforwardly use "axiom" $\boldsymbol{V} = \boldsymbol{v} \Rightarrow \boldsymbol{Y} = \boldsymbol{y} \Box\!\!\rightarrow \varphi' \Leftrightarrow P(\varphi'_{\boldsymbol{Y}=\boldsymbol{y}} \mid \boldsymbol{v}) \geq 1 - \alpha$ to express the "$1 - \alpha$" truth condition of causal counterfactuals.[10] There is another language that is able to express the above statement of the new semantics. Barbero and Virtema have proposed a complete axiomatic system for probabilistic interventionist counterfactuals based on *causal multiteam semantics* [1]. Despite the different semantic framework, we could directly adopt their language to express mixtures of counterfactuals and statements of counterfactual probabilities. More details on Ibeling and Icard's [23] and Barbero and Virtema's [1] work will be discussed in the next section. These two options of extending our language could be referenced in attempts to address the completeness issue, but in this paper, we will not go deep into this problem.

Below I list some typical logical properties for causal counterfactuals, the validity proofs of which are omitted.

Logical properties:		$\models^\alpha$	$\models$
Composition*	$(\mathbf{Y} = \mathbf{y} \Box\!\!\rightarrow X = x) \wedge (\mathbf{Y} = \mathbf{y} \Box\!\!\rightarrow Z = z) \Rightarrow (\mathbf{Y} = \mathbf{y} \wedge X = x) \Box\!\!\rightarrow Z = z$	invalid	valid
Negation (1)	$\mathbf{Y} = \mathbf{y} \Box\!\!\rightarrow \neg\varphi' \Rightarrow \neg\mathbf{Y} = \mathbf{y} \Box\!\!\rightarrow \varphi'$	valid	valid
Negation (2)	$\neg\mathbf{Y} = \mathbf{y} \Box\!\!\rightarrow \varphi' \Rightarrow \mathbf{Y} = \mathbf{y} \Box\!\!\rightarrow \neg\varphi'$	invalid	invalid
Disjunction* (1)	$\mathbf{Y} = \mathbf{y} \Box\!\!\rightarrow (\varphi' \vee \psi') \Rightarrow (\mathbf{Y} = \mathbf{y} \Box\!\!\rightarrow \varphi') \vee (\mathbf{Y} = \mathbf{y} \Box\!\!\rightarrow \psi')$	invalid	invalid
Disjunction* (2)	$(\mathbf{Y} = \mathbf{y} \Box\!\!\rightarrow \varphi') \vee (\mathbf{Y} = \mathbf{y} \Box\!\!\rightarrow \psi') \Rightarrow \mathbf{Y} = \mathbf{y} \Box\!\!\rightarrow (\varphi' \vee \psi')$	valid	valid
Reversibility	$(\mathbf{Y} = \mathbf{y} \wedge X = x) \Box\!\!\rightarrow Z = z \wedge (\mathbf{Y} = \mathbf{y} \wedge Z = z) \Box\!\!\rightarrow X = x \Rightarrow \mathbf{Y} = \mathbf{y} \Box\!\!\rightarrow Z = z$	invalid	valid
Right Weakening	$(\mathbf{Y} = \mathbf{y} \Box\!\!\rightarrow \varphi') \wedge (\varphi' \Rightarrow \psi') \Rightarrow (\mathbf{Y} = \mathbf{y} \Box\!\!\rightarrow \psi')$	valid	valid
Left Strengthening	$(\mathbf{Y} = \mathbf{y} \Box\!\!\rightarrow \mathbf{X} = \mathbf{x}) \wedge (\mathbf{X} = \mathbf{x} \Box\!\!\rightarrow \varphi') \Rightarrow (\mathbf{Y} = \mathbf{y} \Box\!\!\rightarrow \varphi')$	invalid	invalid
K-axiom	$\mathbf{Y} = \mathbf{y} \Box\!\!\rightarrow (\varphi' \Rightarrow \psi') \Rightarrow (\mathbf{Y} = \mathbf{y} \Box\!\!\rightarrow \varphi' \Rightarrow \mathbf{Y} = \mathbf{y} \Box\!\!\rightarrow \psi')$	invalid	valid
Weak Centering	$\mathbf{Y} = \mathbf{y} \Box\!\!\rightarrow \varphi' \Rightarrow (\mathbf{Y} = \mathbf{y} \Rightarrow \varphi')$	valid	valid
Centering	$\mathbf{Y} = \mathbf{y} \wedge \varphi' \Rightarrow \mathbf{Y} = \mathbf{y} \Box\!\!\rightarrow \varphi'$	valid	valid

The combination of Weak Centering and Centering is actually Pearl's Consistency axiom for Pearlian causal models [29, p. 229]. That is, $\mathbf{Y} = \mathbf{y} \Rightarrow (\mathbf{Y} = \mathbf{y} \Box\!\!\rightarrow \varphi') \Leftrightarrow \varphi'$, if the antecedent is true, then the causal counterfactual is equivalent to its consequent.

[10] The expressions of counterfactual probabilities (e.g. $P(\varphi'_{\mathbf{Y}=\boldsymbol{y}} \mid \boldsymbol{v}) \geq 1 - \alpha$) can be rewritten in Ibeling and Icard's language of the third-level causal hierarchy, that is, probabilities for arbitrary boolean combinations of causal counterfactuals.

4 Other Approaches to Probabilistic Causal Logics

In this section, we will compare other literature that incorporates probabilities into the causal modelling framework and proposes relevant logics for counterfactuals.[11]

Ibeling and Icard. Their axiomatization work [23] is the most relevant one in the sense that they also employ the setup of Pearl's probabilistic causal models. The most difference between their probabilistic logics and ours is that they characterize probabilistic counterfactuals, i.e. probabilities of base formulas: (1) sentences without intervention, (2) causal counterfactuals and (3) Boolean combinations of causal counterfactuals, corresponding to different three levels of languages or *causal hierarchy*.[12] The atomic formulas in their language are inequalities of the form $\mathbf{t} \geq \mathbf{t}$ where $\mathbf{t}$ is a polynomial over probabilities of the three kinds of base formulas. Their logics synthesize causal counterfactuals and typical probabilistic logics. Although we both adopt standard probabilistic causal models and the evaluation methodology of probabilistic quantities defined in models, their semantics is not the above-threshold probabilistic conditions for determining the truth of counterfactuals. These differences are revealed via our totally different axiomatic systems. One is more like systems of probabilistic logics, most axioms of which are operation properties of probability instead of characteristic axioms of causal models expressed by counterfactuals.

Beckers. Beckers proposes "nondeterministic causal models" and provides the corresponding sound and complete axiomatization [3]. The causal models are nondeterministic in the sense that the values of the parents determine a range of possible values for a variable rather than a specific value and the assumption of a unique solution after any intervention is dropped. He investigates the probabilistic generalization of nondeterministic causal models, which allows for the exploration of the computation of counterfactual probabilities even in causal Bayesian networks. The generalization is mainly about structural equations, which are transformed into conditional probability distributions $P_X(X|\mathbf{Pa_X})$. The corresponding language he considers is called *probabilistic causal formulas*. The basic probabilistic causal formulas are of the form $(\mathbf{X} = \mathbf{x} \;\Box\!\!\rightarrow \varphi) = p$, where $\mathbf{X} = \mathbf{x} \;\Box\!\!\rightarrow \varphi$ is a causal counterfactual and $p \in [0, 1]$. The truth condition of $(\mathbf{X} = \mathbf{x} \;\Box\!\!\rightarrow \varphi) = p$ relative to a probabilistic nondeterministic causal model M and a solution $(\mathbf{u}, \mathbf{v})$ ($(\mathbf{u}, \mathbf{v})$ is a solution of M if $P_M(\mathbf{u}, \mathbf{v}) > 0$) is that $P_M(\mathbf{X} = \mathbf{x} \;\Box\!\!\rightarrow \varphi|\mathbf{u}, \mathbf{v}) = p$, the evaluation of which in M can be seen

[11] Thank the reviewers for pointing out the necessity of comparisons with other similar approaches and recommending the references.

[12] The causal hierarchy denotes increasing levels of causal knowledge for accomplishing three causal reasoning tasks - prediction, intervention and counterfactuals [29, p. 38]. Syntactically, the prediction/association level corresponds to expressions of conditional probabilities; probabilities of causal counterfactuals constitute the main sentences admitted into the intervention level; the expressions of counterfactual probabilities represent information required in the counterfactual level.

as following Pearl's three-step procedure for computing counterfactual probabilities [29, p. 206]. The semantics is also not a above-threshold probabilistic semantics for counterfactuals, and Beckers doesn't mention axiomatization for this probabilistic logic.

Barbero and Virtema. Barbero and Virtema provide strongly complete axiomatization for probabilistic interventionist counterfactuals [1], which typically consist of operators: marginal probabilities; interventionist counterfactuals; selective implications. The selective implications $\supset$ are capable of representing conditional probabilities so that expressions $P(\alpha|\gamma) \geq \epsilon$ and $P(\alpha|\gamma) \geq P(\beta|\gamma)$ can be defined as $\gamma \supset (P(\alpha) \geq \epsilon)$ and $\gamma \supset (P(\alpha) \geq P(\beta))$ respectively where α, β and γ are formulas in the language of deterministic causal counterfactuals. Therefore, the language of their logic mainly has three kinds of important formulas: (1) causal counterfactuals including nested counterfactuals; (2) probabilistic quantities of the form $P(\alpha) \rhd \epsilon$ and $P(\alpha) \rhd P(\beta)$ where α and β are from (1), $\rhd \in \{\geq, >\}$ and $\epsilon \in [0,1] \cap \mathbb{Q}$; (3) probabilistic counterfactuals of the form $\mathbf{X} = \mathbf{x} \;\Box\!\!\rightarrow \varphi$ in which φ could be any mixture of the formulas in (1), (2) and (3). Their language is rather expressive compared with our simple language of non-probabilistic causal counterfactuals.

The framework they employ is causal multiteam semantics, which combines the features of structural equation modelling and *team semantics*.[13] A *multiteam* can be represented as a table of assignments of all variables with an extra variable *key* that takes different values over different assignments. For example,

Table 1. A multiteam of three binary variables X, Y and Z

key	X	Y	Z
0	0	0	0
1	0	0	1
2	0	1	1

A *causal multiteam* is a pair of a multiteam and a collection of structural equations. The idea of the calculation procedure of various (marginal, conditional, interventional and counterfactual) probabilities in a causal multiteam is analogous to that in Pearl's probabilistic causal models. The difference is that quantities in standard probabilistic causal models would be reduced to the sum of probabilities of contexts (assignments of all exogenous variables), whilst computing probabilities in causal multiteams simply follows the counting measure over multiteams. For instance, in a causal multiteam T with the above table, a

[13] Team semantics is the principal semantics for *dependence logic*. The evaluation of dependence statements in team semantics is associated with sets of variable assignments (*teams*) instead of single assignments [10].

marginal probability $P_T(Y = 0) = \frac{2}{3}$, the ratio between the number of assignments that satisfy $Y = 0$ and the total number of assignments in the table (Table 1).

The truth condition of a probabilistic counterfactual $\mathbf{X} = \mathbf{x} \,\Box\!\!\rightarrow \varphi$ is still that φ is true in the model after the intervention $\mathbf{X} = \mathbf{x}$. To implement an intervention in a multiteam, first we change the values of the variables that are intervened in the multiteam and delete their equations, then calculate and modify the values of other variables in that multiteam. Another critical operation in causal multiteams is observation (or conditionalization), which corresponds to the selective implication or conditional probabilities. "Observing α" is to select assignments that are consistent with α and produces causal sub-multiteams. The resulting axiomatic system mainly consists of characteristic axioms of probabilistic expressions, the selective implication, typical axioms of causal counterfactuals, and the combinations of these expressions.

5 Probabilistic Semantics for Causal Conditionals

One application of probabilistic semantics for causal counterfactuals is to interpret counterfactuals probabilistically in accounts of actual causation in the structural framework, so that we obtain a certain kind of indeterministic accounts for actual causation. Compared with type or general causation, actual (or token, singular) causation is about specific causal relations relative to particular events or facts in particular time and space. With the popularity of the causal modelling approach to causation, there are several proposals of actual causation theories in the structural framework (e.g. [29, chap. 10], [2,11,13,15,17,18,21,22,32,33]). The most influential ones are referred to as the HP (Halpern-Pearl) definitions of actual causation [15,17,18]. I choose the latest HP definition – the modified HP definition [15] (which Halpern prefers [16, p. 27]) to illustrate the new approach to obtaining a probabilistic account of actual causation.

Definition 4 (The Modified HP Definition). *$\boldsymbol{X} = \boldsymbol{x}$ is an actual cause of φ' in a causal setting $\langle M, \boldsymbol{u}\rangle$ if the following three conditions hold:*

AC1. $\langle M, \boldsymbol{u}\rangle \Vdash \boldsymbol{X} = \boldsymbol{x} \wedge \varphi'$.

AC2. There is $\mathbf{W}(\subseteq \mathbf{V})$ and $\boldsymbol{x}'(\in R(\boldsymbol{X}))$ such that if $\langle M, \boldsymbol{u}\rangle \Vdash \boldsymbol{W} = \boldsymbol{w}$, then

$$\langle M, \boldsymbol{u}\rangle \Vdash (\boldsymbol{X} = \boldsymbol{x}' \wedge \boldsymbol{W} = \boldsymbol{w}) \,\Box\!\!\rightarrow \neg\varphi'.$$

AC3. $\boldsymbol{X}$ is minimal; there is no $\boldsymbol{X}_s(\subset \boldsymbol{X})$ such that $\boldsymbol{X}_s = \boldsymbol{x}$ satisfies AC2.

AC1 requires the actuality of a putative cause and the outcome. AC2 is a necessity condition in which the counterfactual dependence between $\boldsymbol{X} = \boldsymbol{x}$ and φ' is evaluated in some actual circumstance $\boldsymbol{W} = \boldsymbol{w}$. AC3 is a minimality condition which ensures that all elements in a cause are necessary. It requires that there cannot be a strict subset of $\boldsymbol{X}$ such that it taking the actual value forms counterfactual dependence with φ'.

We embed the modified HP definition in probabilistic causal models, and obtain its indeterministic version with causal counterfactuals in it interpreted by "equal to 1" semantics:

Definition 5 (C). *$\boldsymbol{X} = \boldsymbol{x}$ is an actual cause of φ' in a probabilistic causal world $\langle M, P, \boldsymbol{u}\rangle$ if and only if*[14]

$\boldsymbol{C_a}$: *$\langle M, P, \boldsymbol{u}\rangle \models \boldsymbol{X} = \boldsymbol{x} \wedge \varphi'$;*

$\boldsymbol{C_b}$: *there is an alternative value $\boldsymbol{x}'$ of $\boldsymbol{X}$ and a set $\boldsymbol{W}$ of endogenous variables whose actual value in $\langle M, P, \boldsymbol{u}\rangle$ is $\boldsymbol{w}$ s.t. $\langle M, P, \boldsymbol{u}\rangle \models (\boldsymbol{X} = \boldsymbol{x}' \wedge \boldsymbol{W} = \boldsymbol{w}) \Box\!\!\rightarrow \neg\varphi'$ (i.e. $P((\neg\varphi')_{\boldsymbol{X}=\boldsymbol{x}',\boldsymbol{W}=\boldsymbol{w}} \mid \boldsymbol{v}(\boldsymbol{u})) = 1$);*

$\boldsymbol{C_c}$: *there is no strict subset $\boldsymbol{X}_s$ of $\boldsymbol{X}$ s.t. $\boldsymbol{X}_s = \boldsymbol{x}$ satisfies $\boldsymbol{C_b}$.*

According to "equal to 1" semantics of causal counterfactuals, $\mathbf{C_b}$ requires that $P((\neg\varphi')_{\boldsymbol{X}=\boldsymbol{x}',\boldsymbol{W}=\boldsymbol{w}} \mid \boldsymbol{v}(\boldsymbol{u})) = 1$ or $P(\varphi'_{\boldsymbol{X}=\boldsymbol{x}',\boldsymbol{W}=\boldsymbol{w}} \mid \boldsymbol{v}(\boldsymbol{u})) = 0$. For "$1-\alpha$" semantics, we can have another probabilistic version of the modified HP definition $\mathbf{C}_\alpha$ which is defined in the same way. Similarly, $\mathbf{C}_{\alpha b}$ actually requires that $P((\neg\varphi')_{\boldsymbol{X}=\boldsymbol{x}',\boldsymbol{W}=\boldsymbol{w}} \mid \boldsymbol{v}(\boldsymbol{u})) \geq 1-\alpha$ or $P(\varphi'_{\boldsymbol{X}=\boldsymbol{x}',\boldsymbol{W}=\boldsymbol{w}} \mid \boldsymbol{v}(\boldsymbol{u})) \leq \alpha$ $(0 \leq \alpha < \frac{1}{2})$.

When it comes to theories of probabilistic causation, a traditional approach is the *probability raising* principle (see [6–9,25,27,30]). Roughly speaking, a cause shall increase the probability of its effect under certain background. If the second conditions in $\mathbf{C}$ and $\mathbf{C}_\alpha$ are written as probability raising inequalities, that is, the probability of φ' would be higher if $\mathbf{X}$ were set to its actual value than if it were set to an alternative value under certain actual circumstances, then the second conditions require that $\mathbf{X}$ raise the probability of φ' significantly as $P(\varphi'_{\boldsymbol{X}=\boldsymbol{x},\boldsymbol{Y}=\boldsymbol{y}} \mid \boldsymbol{v}(\boldsymbol{u})) = 1$ ($\boldsymbol{X} = \boldsymbol{x}$ and $\mathbf{Y} = \mathbf{y}$ are actual events) is far greater than $P(\varphi'_{\boldsymbol{X}=\boldsymbol{x}',\boldsymbol{W}=\boldsymbol{w}} \mid \boldsymbol{v}(\boldsymbol{u})) = 0$ or $P(\varphi'_{\boldsymbol{X}=\boldsymbol{x}',\boldsymbol{W}=\boldsymbol{w}} \mid \boldsymbol{v}(\boldsymbol{u})) \leq \alpha$ $(0 \leq \alpha < \frac{1}{2})$. This requirement is similar to Lewis's probabilistic account of actual causation [27].

In order to see the properties of these probabilistic definitions of actual causation from a logical perspective, we could treat the definitions as semantics for causal claims or causal conditionals and explore the logic. The corresponding languages for the two definitions are similar to $\mathcal{L}(S)$ for causal counterfactuals. A primitive formula is an endogenous variable taking its value. A causal conditional corresponding to definition $\mathbf{C}$ is of the form $\boldsymbol{X} = \boldsymbol{x} \xrightarrow{\mathbf{C}} \varphi'$ in which $\boldsymbol{X} \subseteq \mathbf{V}$, $\boldsymbol{x} \in R(\boldsymbol{X})$ and $\mathbf{X} \neq \emptyset$. It means $\boldsymbol{X} = \boldsymbol{x}$ is an actual cause of φ' under $\mathbf{C}$. The language for $\mathbf{C}$ is Boolean combinations of the corresponding causal conditionals and primitive formulas, denoted as $\mathcal{L}(S)_1$. Likewise, we could have the language $\mathcal{L}(S)_\alpha$ for $\mathbf{C}_\alpha$ in which the causal conditionals are of the form $\boldsymbol{X} = \boldsymbol{x} \xrightarrow{\mathbf{C}_\alpha} \varphi'$.

Actually, the language of causal counterfactuals $\mathcal{L}(S)$ with probabilistic semantics is capable of expressing $\mathbf{C}$, $\mathbf{C}_\alpha$ and any accounts of the significant probability raising of the form $P(\varphi'_{\boldsymbol{X}=\boldsymbol{x},\boldsymbol{Y}_1=\boldsymbol{y}_1} \mid \boldsymbol{v}(\boldsymbol{u})) \geq 1-\alpha > \alpha \geq P(\varphi'_{\boldsymbol{X}=\boldsymbol{x}',\boldsymbol{Y}_2=\boldsymbol{y}_2} \mid \boldsymbol{v}(\boldsymbol{u}))$[15]. Take $\mathbf{C}$ and $\mathbf{C}_\alpha$ as an example, the claim "$\boldsymbol{X} = \boldsymbol{x}$ is

[14] In probabilistic scenarios, instead of determining an actual causal relation, Halpern [16, section 2.5] and Pearl [29, chap. 10] are inclined to talk about probabilities of actual causation. This can be seen as an alternative way to analyze actual causal relations in uncertain circumstances compared with theories of probabilistic causation. Thank one reviewer for raising the question of the difference between probabilistic causation and probability of causation.

[15] When α is 0, $P(\varphi'_{\boldsymbol{X}=\boldsymbol{x},\boldsymbol{Y}_1=\boldsymbol{y}_1} \mid \boldsymbol{v}(\boldsymbol{u})) = 1 > 0 = P(\varphi'_{\boldsymbol{X}=\boldsymbol{x}',\boldsymbol{Y}_2=\boldsymbol{y}_2} \mid \boldsymbol{v}(\boldsymbol{u}))$.

an actual cause of φ'" (or $\boldsymbol{X}=\boldsymbol{x} \xrightarrow{\mathbf{C}} \varphi'$ and $\boldsymbol{X}=\boldsymbol{x} \xrightarrow{\mathbf{C}_\alpha} \varphi'$) can be expressed by the sentence $\boldsymbol{X}=\boldsymbol{x} \wedge \varphi' \wedge \bigvee_{\boldsymbol{W}\subseteq \mathbf{V}, \boldsymbol{x}'\in R(\boldsymbol{X})}(\boldsymbol{W}=\boldsymbol{w} \Rightarrow (\boldsymbol{X}=\boldsymbol{x}' \wedge \boldsymbol{W}=\boldsymbol{w}) \Box\!\!\rightarrow \neg\varphi') \wedge \neg \bigvee_{\boldsymbol{X}_s \subset \boldsymbol{X}} \bigvee_{\boldsymbol{W}'\subseteq \mathbf{V}, \boldsymbol{x}''\in R(\boldsymbol{X}_s)}(\boldsymbol{W}'=\boldsymbol{w}^* \Rightarrow (\boldsymbol{X}_s=\boldsymbol{x}'' \wedge \boldsymbol{W}'=\boldsymbol{w}^*) \Box\!\!\rightarrow \neg\varphi')$.

We define the semantics of $\mathcal{L}(S)_1$ on the set of probabilistic causal models with signature S whose probability functions assign a positive probability to each context.

Definition 6 ($\models_1$). *The satisfaction relation $\models_1$ on φ ($\in \mathcal{L}(S)_1$) and a probabilistic causal world $\langle M, P, \boldsymbol{u}\rangle$ is defined as follows.*

$\langle M,P,\boldsymbol{u}\rangle \models_1 X=x$	$\iff$	$\langle M,\boldsymbol{u}\rangle \Vdash X=x$
$\langle M,P,\boldsymbol{u}\rangle \models_1 \boldsymbol{X}=\boldsymbol{x} \xrightarrow{\mathbf{C}} \varphi'$	$\iff$	$\boldsymbol{X}=\boldsymbol{x}$ is an actual cause of φ' under $\mathbf{C}$
$\langle M,P,\boldsymbol{u}\rangle \models_1 \neg\varphi$	$\iff$	$\langle M,P,\boldsymbol{u}\rangle \not\models_1 \varphi$
$\langle M,P,\boldsymbol{u}\rangle \models_1 (\varphi \wedge \psi)$	$\iff$	$\langle M,P,\boldsymbol{u}\rangle \models_1 \varphi$ and $\langle M,P,\boldsymbol{u}\rangle \models_1 \psi$

A formula φ is valid on $\langle M, P\rangle$ if and only if for any $\boldsymbol{u}$, $\langle M,P,\boldsymbol{u}\rangle \models_1 \varphi$.

The semantics with respect to $\mathbf{C}_\alpha$ is defined in the same way. There is no difference between $\mathbf{C}$ and $\mathbf{C}_\alpha$ in the causal properties presented below. Thus, I simply use $\mathbf{C}$ to represent $\mathbf{C}$ and $\mathbf{C}_\alpha$ in causal conditionals hereinafter. The validity and invalidity of the following properties are easy to verify, so the proofs are omitted.

Valid causal properties for C and $\mathbf{C}_\alpha$:

(1) Actuality $(\boldsymbol{X}=\boldsymbol{x} \xrightarrow{\mathbf{C}} \varphi') \Rightarrow (\boldsymbol{X}=\boldsymbol{x} \wedge \varphi')$
This property is guaranteed by the first condition of $\mathbf{C}$ and $\mathbf{C}_\alpha$.

(2) Minimality $(\boldsymbol{X}=\boldsymbol{x} \xrightarrow{\mathbf{C}} \varphi') \Rightarrow (\neg \bigvee_{\boldsymbol{X}_s \subset \boldsymbol{X}} \boldsymbol{X}_s=\boldsymbol{x} \xrightarrow{\mathbf{C}} \varphi')$
This claim corresponds to the third conditions of $\mathbf{C}$ and $\mathbf{C}_\alpha$. It follows that actual causes cannot be cut or extended, that is $((\boldsymbol{X}=\boldsymbol{x} \wedge \boldsymbol{Y}=\boldsymbol{y}) \xrightarrow{\mathbf{C}} \varphi') \Rightarrow (\neg \boldsymbol{X}=\boldsymbol{x} \xrightarrow{\mathbf{C}} \varphi')$ and $(\boldsymbol{X}=\boldsymbol{x} \xrightarrow{\mathbf{C}} \varphi') \Rightarrow (\neg(\boldsymbol{X}=\boldsymbol{x} \wedge \boldsymbol{Y}=\boldsymbol{y}) \xrightarrow{\mathbf{C}} \varphi')$ are valid.

(3) Distribution $((\boldsymbol{X}=\boldsymbol{x} \xrightarrow{\mathbf{C}} (\varphi' \Rightarrow \psi')) \wedge \varphi') \Rightarrow (\bigvee_{\boldsymbol{X}_s \subseteq \boldsymbol{X}} \boldsymbol{X}_s=\boldsymbol{x} \xrightarrow{\mathbf{C}} \psi')$
This claim can be instantiated by the case "*if* that it rained and I didn't take an umbrella caused that if I had gone out, then I would have been caught in the rain and I actually went out, *then* that it rained caused that I was caught in the rain, or that I didn't take an umbrella caused that I was caught in the rain, or the two factors together caused that I was caught in the rain." This property says that when the outcome is an implication, the consequent of that implication can be an outcome of the cause.

(4) Union $((\boldsymbol{X}=\boldsymbol{x} \xrightarrow{\mathbf{C}} \varphi') \wedge \psi') \Rightarrow (\bigvee_{\boldsymbol{X}_s \subseteq \boldsymbol{X}} \boldsymbol{X}_s=\boldsymbol{x} \xrightarrow{\mathbf{C}} (\varphi' \wedge \psi'))$
"If that it rained caused that the ground was wet, and I went out, then that it rained caused that the ground was wet and that I went out." It seems strange to add any actual event in the outcome. In fact, all definitions based on the counterfactual rationale have this problem. Suppose we know that A causes B, then if it were not A, B would not happen. Then if it were not A,

then B and C would not happen. We could count A as a cause of B and C. Thus for the definitions of causation we've presented, the cause part cannot be extended because of the minimality condition, but the outcome can be extended.

(5) Or $(\boldsymbol{X} = \boldsymbol{x} \xrightarrow{\mathbf{C}} (\varphi' \vee \psi')) \Rightarrow (\bigvee_{\boldsymbol{X}_s \subseteq \boldsymbol{X}} \boldsymbol{X}_s = \boldsymbol{x} \xrightarrow{\mathbf{C}} \varphi' \vee \bigvee_{\boldsymbol{X}_s \subseteq \boldsymbol{X}} \boldsymbol{X}_s = \boldsymbol{x} \xrightarrow{\mathbf{C}} \psi')$
An example of this claim could be: if that it rained caused that the ground was wet or that I was caught in the rain, then that it rained caused that the ground was wet or that it rained caused that I was caught in the rain. Which causal relation holds in the consequent only depends on which disjunct φ' or ψ' is actual. Thus we have the valid claim $((\boldsymbol{X} = \boldsymbol{x} \xrightarrow{\mathbf{C}} (\varphi' \vee \psi')) \wedge \varphi') \Rightarrow (\bigvee_{\boldsymbol{X}_s \subseteq \boldsymbol{X}} \boldsymbol{X}_s = \boldsymbol{x} \xrightarrow{\mathbf{C}} \varphi')$.

(6) Consequent Equivalence *From* $\varphi' \Leftrightarrow \psi'$, *infer* $\boldsymbol{X} = \boldsymbol{x} \xrightarrow{\mathbf{C}} \varphi' \Leftrightarrow \boldsymbol{X} = \boldsymbol{x} \xrightarrow{\mathbf{C}} \psi'$

This rule implies that a causal claim would still hold if we change the logical form of the outcome sentence.

Invalid logical properties for C and $\mathbf{C}_\alpha$:

(1) Conjunction $(\boldsymbol{X} = \boldsymbol{x} \xrightarrow{\mathbf{C}} \varphi' \wedge \boldsymbol{X} = \boldsymbol{x} \xrightarrow{\mathbf{C}} \psi') \Rightarrow (\boldsymbol{X} = \boldsymbol{x} \xrightarrow{\mathbf{C}} (\varphi' \wedge \psi'))$
The valid claim is $(\boldsymbol{X} = \boldsymbol{x} \xrightarrow{\mathbf{C}} \varphi' \wedge \boldsymbol{X} = \boldsymbol{x} \xrightarrow{\mathbf{C}} \psi') \Rightarrow (\bigvee_{\boldsymbol{X}_s \subseteq \boldsymbol{X}} \boldsymbol{X}_s = \boldsymbol{x} \xrightarrow{\mathbf{C}} (\varphi' \wedge \psi'))$ and $((\boldsymbol{X} = \boldsymbol{x} \xrightarrow{\mathbf{C}} \varphi') \wedge \psi') \Rightarrow (\bigvee_{\boldsymbol{X}_s \subseteq \boldsymbol{X}} \boldsymbol{X}_s = \boldsymbol{x} \xrightarrow{\mathbf{C}} (\varphi' \wedge \psi'))$ (Union).

(2) Contraction $(\boldsymbol{X} = \boldsymbol{x} \xrightarrow{\mathbf{C}} (\varphi' \wedge \psi')) \Rightarrow (\boldsymbol{X} = \boldsymbol{x} \xrightarrow{\mathbf{C}} \varphi' \vee \boldsymbol{X} = \boldsymbol{x} \xrightarrow{\mathbf{C}} \psi')$
There can be an example showing that: if the cause event $\boldsymbol{X} = \boldsymbol{x}$ didn't occur, effect A/B would not always or in most occasions be absent, which means the former cannot be judged as a cause of the latter; however, effect $A \wedge B$ would always or in most occasions be absent if $\boldsymbol{X} = \boldsymbol{x}$ did not show up, which means the conjunction of the effects instead could be caused by $\boldsymbol{X} = \boldsymbol{x}$ according to $\mathbf{C}$ and $\mathbf{C}_\alpha$.

(3) Transitivity $(\boldsymbol{X} = \boldsymbol{x} \xrightarrow{\mathbf{C}} \boldsymbol{Y} = \boldsymbol{y} \wedge \boldsymbol{Y} = \boldsymbol{y} \xrightarrow{\mathbf{C}} \varphi') \Rightarrow (\boldsymbol{X} = \boldsymbol{x} \xrightarrow{\mathbf{C}} \varphi')$
The reason why Transitivity is invalid is that the scenario where $\boldsymbol{X}$ can make $\boldsymbol{Y} = \boldsymbol{y}$ false may not be the scenario where φ' would change.

(4) Right Weakening $((\boldsymbol{X} = \boldsymbol{x} \xrightarrow{\mathbf{C}} \varphi') \wedge (\varphi' \Rightarrow \psi')) \Rightarrow (\boldsymbol{X} = \boldsymbol{x} \xrightarrow{\mathbf{C}} \psi')$
The other version of this claim is: *From* $\varphi' \Rightarrow \psi'$, *infer* $(\boldsymbol{X} = \boldsymbol{x} \xrightarrow{\mathbf{C}} \varphi') \Rightarrow (\boldsymbol{X} = \boldsymbol{x} \xrightarrow{\mathbf{C}} \psi')$. They are invalid mainly because ψ' may not change in the scenario where $\boldsymbol{X}$ changes φ'. But we do have the valid rule Consequent Equivalence: *From* $\varphi' \Leftrightarrow \psi'$, *infer* $\boldsymbol{X} = \boldsymbol{x} \xrightarrow{\mathbf{C}} \varphi' \Leftrightarrow \boldsymbol{X} = \boldsymbol{x} \xrightarrow{\mathbf{C}} \psi'$.

(5) Left Strengthening $((\boldsymbol{X} = \boldsymbol{x} \Rightarrow \boldsymbol{Y} = \boldsymbol{y}) \wedge (\boldsymbol{Y} = \boldsymbol{y} \xrightarrow{\mathbf{C}} \varphi') \wedge \boldsymbol{X} = \boldsymbol{x}) \Rightarrow (\boldsymbol{X} = \boldsymbol{x} \xrightarrow{\mathbf{C}} \varphi')$
A counterexample of this claim could be: if Bob's black teeth implies his

heavy smoking, that he smoked heavily caused his lung cancer and he actually has black teeth, then Bob's black teeth caused his lung cancer. The other version of this property is that *From* $\boldsymbol{X}=\boldsymbol{x} \Rightarrow \boldsymbol{Y}=\boldsymbol{y}$, *infer* $((\boldsymbol{Y}=\boldsymbol{y} \xrightarrow{\mathbf{C}} \varphi') \wedge \boldsymbol{X}=\boldsymbol{x}) \Rightarrow (\boldsymbol{X}=\boldsymbol{x} \xrightarrow{\mathbf{C}} \varphi')$. They are invalid mainly because $\boldsymbol{X}$ may not create the scenario where $\boldsymbol{Y}$ makes a difference to φ' by changing its value so that we have the counterfactual relation between $\boldsymbol{X}$ and φ'.

(6) Disjunction $(\boldsymbol{X}=\boldsymbol{x} \xrightarrow{\mathbf{C}} \varphi' \vee \boldsymbol{X}=\boldsymbol{x} \xrightarrow{\mathbf{C}} \psi') \Rightarrow (\boldsymbol{X}=\boldsymbol{x} \xrightarrow{\mathbf{C}} (\varphi' \vee \psi'))$
The reason why it is invalid is that $\boldsymbol{X}$ could make a difference to φ' in some situations where ψ' may not change. We can write this property as $(\boldsymbol{X}=\boldsymbol{x} \xrightarrow{\mathbf{C}} \varphi') \Rightarrow (\boldsymbol{X}=\boldsymbol{x} \xrightarrow{\mathbf{C}} (\varphi' \vee \psi'))$. Another similar claim is $(\boldsymbol{X}=\boldsymbol{x} \xrightarrow{\mathbf{C}} \varphi' \wedge \boldsymbol{X}=\boldsymbol{x} \xrightarrow{\mathbf{C}} \psi') \Rightarrow (\boldsymbol{X}=\boldsymbol{x} \xrightarrow{\mathbf{C}} (\varphi' \vee \psi'))$ which is invalid for the same reason.

(7) Separation $(\boldsymbol{X}=\boldsymbol{x} \xrightarrow{\mathbf{C}} (\varphi' \wedge \psi')) \Rightarrow (\boldsymbol{X}=\boldsymbol{x} \xrightarrow{\mathbf{C}} \varphi' \wedge \boldsymbol{X}=\boldsymbol{x} \xrightarrow{\mathbf{C}} \psi')$
It is invalid because the counterfactual relation between $\boldsymbol{X}=\mathbf{x}$ and $\varphi' \wedge \psi'$ cannot guarantee the counterfactual dependence between $\boldsymbol{X}$ and one of φ' and ψ'.

(8) Cut $(\boldsymbol{X}=\boldsymbol{x} \xrightarrow{\mathbf{C}} \boldsymbol{Y}=\boldsymbol{y} \wedge (\boldsymbol{X}=\boldsymbol{x} \wedge \boldsymbol{Y}=\boldsymbol{y}) \xrightarrow{\mathbf{C}} \varphi') \Rightarrow (\boldsymbol{X}=\boldsymbol{x} \xrightarrow{\mathbf{C}} \varphi')$
Cut is invalid because of the minimality condition.

For the semantics based on the deterministic modified HP definition, the above valid causal properties for $\mathbf{C}$ and $\mathbf{C}_\alpha$ are all valid, and the invalid properties except Conjunction and Contraction are also invalid. For Contraction, it seems natural to imply "the throwing of the rock caused the window to break or it caused the vase to break" from "the throwing of the rock caused the window and the vase to break." The invalidity of Contraction to some extent indicates that $\mathbf{C}$ and $\mathbf{C}_\alpha$ underperform in exhibiting ordinary causal thoughts.

If we interpret causal counterfactuals probabilistically in some other deterministic definitions of actual causation with sufficiency and necessity conditions, we could also obtain their indeterministic versions of significant probability raising. For the indeterministic version of the modified HP definition, $\mathbf{C}$ with "equal to 1" semantics actually requires $P((\neg\varphi')_{\boldsymbol{X}=\boldsymbol{x}',\boldsymbol{W}=\boldsymbol{w}} \mid \boldsymbol{v}(\boldsymbol{u}))=1$ or $P(\varphi'_{\boldsymbol{X}=\boldsymbol{x}',\boldsymbol{W}=\boldsymbol{w}} \mid \boldsymbol{v}(\boldsymbol{u}))=0$ in the counterfactual condition. Similarly, the probability interpretation of $(\boldsymbol{X}=\boldsymbol{x}' \wedge \boldsymbol{W}=\boldsymbol{w}) \mathrel{\Box\!\!\rightarrow} \neg\varphi'$ in the other indeterministic version $\mathbf{C}_\alpha$ is $P((\neg\varphi')_{\boldsymbol{X}=\boldsymbol{x}',\boldsymbol{W}=\boldsymbol{w}} \mid \boldsymbol{v}(\boldsymbol{u})) \geq 1-\alpha$ $(0 \leq \alpha < \frac{1}{2})$ or $P(\varphi'_{\boldsymbol{X}=\boldsymbol{x}',\boldsymbol{W}=\boldsymbol{w}} \mid \boldsymbol{v}(\boldsymbol{u})) \leq \alpha$. The other sides of probability raising for $P(\varphi'_{\boldsymbol{X}=\boldsymbol{x}',\boldsymbol{W}=\boldsymbol{w}} \mid \boldsymbol{v}(\boldsymbol{u}))=0$ and $P(\varphi'_{\boldsymbol{X}=\boldsymbol{x}',\boldsymbol{W}=\boldsymbol{w}} \mid \boldsymbol{v}(\boldsymbol{u})) \leq \alpha$ are of the form $P(\varphi'_{\boldsymbol{X}=\boldsymbol{x},\boldsymbol{Y}=\boldsymbol{y}} \mid \boldsymbol{v}(\boldsymbol{u}))=1$ ($\boldsymbol{X}=\boldsymbol{x}$ and $\mathbf{Y}=\mathbf{y}$ are actual events), then that means $\boldsymbol{X}=\boldsymbol{x}$ makes a dramatic difference to φ' in $\mathbf{C}$ and $\mathbf{C}_\alpha$, in other words, $\boldsymbol{X}=\boldsymbol{x}$ raises the probability of φ' significantly in $\mathbf{C}$ and $\mathbf{C}_\alpha$. When it comes to the original [17] and updated [18] HP definitions, apart from the necessity condition $\langle M,\boldsymbol{u}\rangle \Vdash (\boldsymbol{X}=\boldsymbol{x}' \wedge \boldsymbol{W}=\boldsymbol{w}') \mathrel{\Box\!\!\rightarrow} \neg\varphi'$ ($\boldsymbol{w}'$ is a setting of $\boldsymbol{W}$), they also have the sufficiency conditions $\langle M, \boldsymbol{u}\rangle \Vdash (\boldsymbol{X} = \boldsymbol{x} \wedge \boldsymbol{W} = \boldsymbol{w}' \wedge \boldsymbol{Z}_s = \boldsymbol{z}) \,\Box\!\!\rightarrow \varphi'$[16] (for the original HP definition) and $\langle M, \boldsymbol{u}\rangle \Vdash (\boldsymbol{X} = \boldsymbol{x} \wedge \boldsymbol{W}_s = \boldsymbol{w}' \wedge \boldsymbol{Z}_s = \boldsymbol{z}) \,\Box\!\!\rightarrow \varphi'$[17] (for the updated HP definition). Once these causal counterfactuals are interpreted by "equal to 1" and "$1-\alpha$" semantics, we have $P(\varphi'_{\boldsymbol{X}=\boldsymbol{x},\boldsymbol{W}=\boldsymbol{w}',\boldsymbol{Z}_s=\boldsymbol{z}} \mid \boldsymbol{v}(\boldsymbol{u})) = 1 > 0 = P(\varphi'_{\boldsymbol{X}=\boldsymbol{x}',\boldsymbol{W}=\boldsymbol{w}'} \mid \boldsymbol{v}(\boldsymbol{u}))$ and $P(\varphi'_{\boldsymbol{X}=\boldsymbol{x},\boldsymbol{W}=\boldsymbol{w}',\boldsymbol{Z}_s=\boldsymbol{z}} \mid \boldsymbol{v}(\boldsymbol{u})) \geq 1-\alpha > \alpha \geq P(\varphi'_{\boldsymbol{X}=\boldsymbol{x}',\boldsymbol{W}=\boldsymbol{w}'} \mid \boldsymbol{v}(\boldsymbol{u}))$ for the original HP definition. It is clear that the indeterministic versions of the original HP definition require significant probability increase.[18] Similarly, the indeterministic versions of the updated HP definition have the inequalities of significant probability raising. Thus, for a general counterfactual account of actual causation with sufficient and necessary conditions like the above two definitions, its indeterministic version would have dramatic probability increase once counterfactuals are interpreted as their corresponding probability quantities being very high. This new approach derives accounts with significant probability raising, while following the traditional probability raising approach, we can also have an account with that feature, e.g. Lewis's theory of probabilistic actual causation [27]. The difference shall be that there always exist the thresholds from the probabilistic semantics of counterfactuals in the inequalities for the new approach, whereas Lewis requires that the chance of the outcome if a putative cause hadn't occurred would have been very much less than the chance of the outcome at the time when the cause occurred. In this paper, I use logical properties to better understand the new probabilistic accounts rather than examining them via typical causal examples. Thus, the analysis of the new approach of probabilistic actual causation is mainly from logical perspective instead of focusing on practical or useful evaluation, which may be investigated in future work.

6 Conclusion

In this paper, I propose probabilistic semantics for causal counterfactuals following the spirit of Leitgeb [26]'s probabilistic semantics for counterfactuals, and provide a complete axiomatic system for "equal to 1" semantics. Specifically, causal counterfactuals are interpreted in probabilistic causal models by virtue of the relevant counterfactual probabilities being sufficiently high. One application of those semantics is to interpret causal counterfactuals in existing counterfactual definitions of actual causation so that we obtain the indeterministic accounts of actual causation. The new probabilistic accounts require significant probability raising between a putative cause and the outcome, and are treated as the semantics of causal conditionals, the causal or conditional properties of which are discussed to better understand the new accounts.

16 $\boldsymbol{w}'$ is the same value setting used in the necessity condition. $(\boldsymbol{W}, \boldsymbol{Z})$ is a partition of all the endogenous variables. $\boldsymbol{z}$ is the actual value setting of $\boldsymbol{Z}$ and $\boldsymbol{Z}_s$ is any subset of $\boldsymbol{Z}$.

17 $\boldsymbol{W}_s$ is any subset of $\boldsymbol{W}$. $(\boldsymbol{W}, \boldsymbol{Z})$ is a partition of all the endogenous variables and $\boldsymbol{Z}_s$ is any subset of $\boldsymbol{Z}$.

18 For "$1-\alpha$" version, the closer α is to 0, the more difference $\boldsymbol{X} = \boldsymbol{x}$ makes to φ'.

There are several directions that are worthwhile to be explored further. The first is the generalization of probabilistic causal models. Note that probabilistic causal models involve Pearlian models which have the property "unique solution" [29, p. 205]. Structural equation models could have multiple solutions in a context before and after intervention. Therefore, Pearl's probabilistic causal models could be generalized to embed causal models with multiple solutions. The tricky part of this idea is the probability redistribution $P_{\boldsymbol{Y}=\boldsymbol{y}}(\boldsymbol{u},\boldsymbol{v})$ where $\mathbf{Y}=\mathbf{y}$ is an intervention. In these generalized probabilistic causal models, we can still explore the corresponding semantics for causal counterfactuals and its axiomatization.

The second possible direction is to extend the language for "$1-\alpha$" semantics so as to complete its axiomatization, for this semantics requires probability actually play a role compared with "equal to 1" semantics. We could consider adding the expressions of probability functions in the language (cf. [1]) so that the distinctive properties of causal counterfactuals with "$1-\alpha$" semantics could be better understood.

Acknowledgments. I'd like to thank Jiji Zhang and Peter Hawke for their discussions and suggestions on this research, and Luke Fenton-Glynn and Rafael De Clercq for their inspiring comments on the earlier manuscript. Thanks to the anonymous reviewers for the insightful comments, and the participants in the TLLM workshop for raising valuable questions. This paper was supported by the key project 23 & ZD240 of National Office for Philosophy and Social Sciences.

Appendix

Lemma 1. *For the following two theorems of* ***SPCL(S)****, let* $\mathbf{Y}\cup\{X_1,...,X_m\}=\mathbf{V}$, $\mathbf{Y}\cap\{X_1,...,X_m\}=\emptyset$, $\mathbf{y}$ *and* x_i *be the value settings of* $\mathbf{Y}$ *and* X_i *in* $\mathbf{v}$ *respectively.*

1. *SOLUTION':* $\vdash (\mathbf{U}=\mathbf{u}\wedge\mathbf{Y}=\mathbf{y})\ \Box\!\!\rightarrow \mathbf{V}=\mathbf{v} \Rightarrow (\mathbf{U}=\mathbf{u}\wedge\mathbf{Y}=\mathbf{y}\wedge\mathbf{V}_{\mathbf{Y}\cup\{X_1\}}=\mathbf{v}_{\mathbf{Y}\cup\{X_1\}})\ \Box\!\!\rightarrow X_1=x_1\wedge...\wedge(\mathbf{U}=\mathbf{u}\wedge\mathbf{Y}=\mathbf{y}\wedge\mathbf{V}_{\mathbf{Y}\cup\{X_m\}}=\mathbf{v}_{\mathbf{Y}\cup\{X_m\}})\ \Box\!\!\rightarrow X_m=x_m$.
2. *POSSIBILITY:* $\vdash \boldsymbol{Y}=\boldsymbol{y}\ \Box\!\!\rightarrow \boldsymbol{V}=\boldsymbol{v} \Rightarrow \bigvee_{\boldsymbol{u}\in R(\boldsymbol{U})}((\boldsymbol{U}=\boldsymbol{u}\wedge\boldsymbol{Y}=\boldsymbol{y}\wedge\boldsymbol{V}_{\boldsymbol{Y}\cup\{X_1\}}=\boldsymbol{v}_{\boldsymbol{Y}\cup\{X_1\}})\ \Box\!\!\rightarrow X_1=x_1\wedge...\wedge(\boldsymbol{U}=\boldsymbol{u}\wedge\boldsymbol{Y}=\boldsymbol{y}\wedge\boldsymbol{V}_{\boldsymbol{Y}\cup\{X_m\}}=\boldsymbol{v}_{\boldsymbol{Y}\cup\{X_m\}})\ \Box\!\!\rightarrow X_m=x_m)$.

Proof. 1. By applying COMPOSITION, we could move $\mathbf{V}_{\mathbf{Y}\cup\{X_1\}}=\mathbf{v}_{\mathbf{Y}\cup\{X_1\}}$ from $\mathbf{V}=\mathbf{v}$ to the antecedent of $(\mathbf{U}=\mathbf{u}\wedge\mathbf{Y}=\mathbf{y})\ \Box\!\!\rightarrow \mathbf{V}=\mathbf{v}$ step by step. We have $(\mathbf{U}=\mathbf{u}\wedge\mathbf{Y}=\mathbf{y})\ \Box\!\!\rightarrow \mathbf{V}=\mathbf{v}\Rightarrow(\mathbf{U}=\mathbf{u}\wedge\mathbf{Y}=\mathbf{y}\wedge X_2=x_2)\ \Box\!\!\rightarrow \mathbf{V}_{X_2}=\mathbf{v}_{X_2}$, $(\mathbf{U}=\mathbf{u}\wedge\mathbf{Y}=\mathbf{y}\wedge X_2=x_2)\ \Box\!\!\rightarrow \mathbf{V}_{X_2}=\mathbf{v}_{X_2}\Rightarrow(\mathbf{U}=\mathbf{u}\wedge\mathbf{Y}=\mathbf{y}\wedge X_2=x_2\wedge X_3=x_3)\ \Box\!\!\rightarrow \mathbf{V}_{X_2,X_3}=\mathbf{v}_{X_2,X_3}$. It follows that $(\mathbf{U}=\mathbf{u}\wedge\mathbf{Y}=\mathbf{y})\ \Box\!\!\rightarrow \mathbf{V}=\mathbf{v}\Rightarrow(\mathbf{U}=\mathbf{u}\wedge\mathbf{Y}=\mathbf{y}\wedge X_2=x_2\wedge X_3=x_3)\ \Box\!\!\rightarrow \mathbf{V}_{X_2,X_3}=\mathbf{v}_{X_2,X_3}$ according to the Syllogistic rule. Repeat the argument, we obtain $(\mathbf{U}=\mathbf{u}\wedge\mathbf{Y}=\mathbf{y})\ \Box\!\!\rightarrow \mathbf{V}=\mathbf{v}\Rightarrow(\mathbf{U}=\mathbf{u}\wedge\mathbf{Y}=\mathbf{y}\wedge\mathbf{V}_{\mathbf{Y}\cup\{X_1\}}=\mathbf{v}_{\mathbf{Y}\cup\{X_1\}})\ \Box\!\!\rightarrow X_1=x_1$. Likewise, we have $(\mathbf{U}=\mathbf{u}\wedge\mathbf{Y}=\mathbf{y})\ \Box\!\!\rightarrow \mathbf{V}=\mathbf{v}\Rightarrow(\mathbf{U}=\mathbf{u}\wedge\mathbf{Y}=\mathbf{y}\wedge\mathbf{V}_{\mathbf{Y}\cup\{X_2\}}=\mathbf{v}_{\mathbf{Y}\cup\{X_2\}})\ \Box\!\!\rightarrow X_2=x_2$,..., $(\mathbf{U}=\mathbf{u}\wedge\mathbf{Y}=\mathbf{y})\ \Box\!\!\rightarrow$

$\mathbf{V} = \mathbf{v} \Rightarrow (\mathbf{U} = \mathbf{u} \wedge \mathbf{Y} = \mathbf{y} \wedge \mathbf{V}_{\mathbf{Y}\cup\{X_m\}} = \mathbf{v}_{\mathbf{Y}\cup\{X_m\}}) \Box\!\!\rightarrow X_m = x_m$. It follows that $(\mathbf{U} = \mathbf{u} \wedge \mathbf{Y} = \mathbf{y}) \Box\!\!\rightarrow \mathbf{V} = \mathbf{v} \Rightarrow (\mathbf{U} = \mathbf{u} \wedge \mathbf{Y} = \mathbf{y} \wedge \mathbf{V}_{\mathbf{Y}\cup\{X_1\}} = \mathbf{v}_{\mathbf{Y}\cup\{X_1\}}) \Box\!\!\rightarrow X_1 = x_1 \wedge ... \wedge (\mathbf{U} = \mathbf{u} \wedge \mathbf{Y} = \mathbf{y} \wedge \mathbf{V}_{\mathbf{Y}\cup\{X_m\}} = \mathbf{v}_{\mathbf{Y}\cup\{X_m\}}) \Box\!\!\rightarrow X_m = x_m$.

2. According to DEFINITENESS, each endogenous variable would take on a certain value, so there always exists an "actual" solution among $\bigvee_{\mathbf{v_i}\in R(\mathbf{V})} \mathbf{V} = \mathbf{v_i}$. For each $\boldsymbol{V} = \boldsymbol{v}_i$, we have an instance of TRANSFORMATION, $\mathbf{V} = \mathbf{v_i} \Rightarrow \mathbf{Y} = \mathbf{y} \Box\!\!\rightarrow \boldsymbol{V} = \boldsymbol{v} \Leftrightarrow \bigwedge_{\mathbf{u}\in R(\mathbf{U})} (\mathbf{U} = \mathbf{u} \Box\!\!\rightarrow \mathbf{V} = \mathbf{v_i} \Rightarrow (\mathbf{U} = \mathbf{u} \wedge \mathbf{Y} = \mathbf{y}) \Box\!\!\rightarrow \boldsymbol{V} = \boldsymbol{v})$. Then $\mathbf{V} = \mathbf{v_i} \Rightarrow \mathbf{Y} = \mathbf{y} \Box\!\!\rightarrow \boldsymbol{V} = \boldsymbol{v} \Rightarrow \bigwedge_{\mathbf{u}\in R(\mathbf{U})} (\mathbf{U} = \mathbf{u} \Box\!\!\rightarrow \mathbf{V} = \mathbf{v_i} \Rightarrow (\mathbf{U} = \mathbf{u} \wedge \mathbf{Y} = \mathbf{y}) \Box\!\!\rightarrow \boldsymbol{V} = \boldsymbol{v})$ is a theorem. By applying the rule of substitution of equivalents and SOLUTION*, we have the theorem

$$\begin{aligned}\mathbf{V} = \mathbf{v_i} \Rightarrow \mathbf{Y} = \mathbf{y} \Box\!\!\rightarrow \boldsymbol{V} = \boldsymbol{v} \Rightarrow \bigwedge_{\mathbf{u}\in R(\mathbf{U})} (\mathbf{U} = \mathbf{u} \Box\!\!\rightarrow \mathbf{V} = \mathbf{v_i} \Rightarrow \\ (\mathbf{U} = \mathbf{u} \wedge \mathbf{Y} = \mathbf{y} \wedge \mathbf{V}_{\mathbf{Y}\cup\{X_1\}} = \mathbf{v}_{\mathbf{Y}\cup\{X_1\}}) \Box\!\!\rightarrow X_1 = x_1 \wedge ... \wedge \\ (\mathbf{U} = \mathbf{u} \wedge \mathbf{Y} = \mathbf{y} \wedge \mathbf{V}_{\mathbf{Y}\cup\{X_m\}} = \mathbf{v}_{\mathbf{Y}\cup\{X_m\}}) \Box\!\!\rightarrow X_m = x_m).\end{aligned} \tag{1}$$

An instance of CAUSAL WORLD is $\boldsymbol{V} = \boldsymbol{v}_i \Rightarrow \bigvee_{u\in R(\mathbf{U})} \mathbf{U} = \mathbf{u} \Box\!\!\rightarrow \mathbf{V} = \mathbf{v_i}$. By using the rules in propositional logic, we could derive

$$\boldsymbol{V} = \boldsymbol{v}_i \Rightarrow \mathbf{Y} = \mathbf{y} \Box\!\!\rightarrow \boldsymbol{V} = \boldsymbol{v} \Rightarrow \bigvee_{u\in R(\mathbf{U})} \mathbf{U} = \mathbf{u} \Box\!\!\rightarrow \mathbf{V} = \mathbf{v_i}, \tag{2}$$

then combine the consequents of (1) and (2), and obtain (3):

$$\begin{aligned}\mathbf{V} = \mathbf{v_i} \Rightarrow \mathbf{Y} = \mathbf{y} \Box\!\!\rightarrow \boldsymbol{V} = \boldsymbol{v} \Rightarrow \bigvee_{u\in R(\mathbf{U})} (\mathbf{U} = \mathbf{u} \Box\!\!\rightarrow \mathbf{V} = \mathbf{v_i}) \wedge \bigwedge_{\mathbf{u}\in R(\mathbf{U})} \\ (\mathbf{U} = \mathbf{u} \Box\!\!\rightarrow \mathbf{V} = \mathbf{v_i} \Rightarrow (\mathbf{U} = \mathbf{u} \wedge \mathbf{Y} = \mathbf{y} \wedge \mathbf{V}_{\mathbf{Y}\cup\{X_1\}} = \mathbf{v}_{\mathbf{Y}\cup\{X_1\}}) \Box\!\!\rightarrow \\ X_1 = x_1 \wedge ... \wedge (\mathbf{U} = \mathbf{u} \wedge \mathbf{Y} = \mathbf{y} \wedge \mathbf{V}_{\mathbf{Y}\cup\{X_m\}} = \mathbf{v}_{\mathbf{Y}\cup\{X_m\}}) \Box\!\!\rightarrow X_m = x_m).\end{aligned} \tag{3}$$

I write $(\mathbf{U} = \mathbf{u_j} \wedge \mathbf{Y} = \mathbf{y} \wedge \mathbf{V}_{\mathbf{Y}\cup\{X_1\}} = \mathbf{v}_{\mathbf{Y}\cup\{X_1\}}) \Box\!\!\rightarrow X_1 = x_1 \wedge ... \wedge (\mathbf{U} = \mathbf{u_j} \wedge \mathbf{Y} = \mathbf{y} \wedge \mathbf{V}_{\mathbf{Y}\cup\{X_m\}} = \mathbf{v}_{\mathbf{Y}\cup\{X_m\}}) \Box\!\!\rightarrow X_m = x_m$ as $Sol(\mathbf{u_j})$, then the consequent of (3) can be sorted out as:

$$\begin{aligned}((\mathbf{U} = \mathbf{u_1} \Box\!\!\rightarrow \mathbf{V} = \mathbf{v_i}) \wedge (\mathbf{U} = \mathbf{u_1} \Box\!\!\rightarrow \mathbf{V} = \mathbf{v_i} \Rightarrow Sol(\mathbf{u_1})) \wedge (\mathbf{U} = \mathbf{u_2} \Box\!\!\rightarrow \\ \mathbf{V} = \mathbf{v_i} \Rightarrow Sol(\mathbf{u_2})) \wedge ... \wedge (\mathbf{U} = \mathbf{u_k} \Box\!\!\rightarrow \mathbf{V} = \mathbf{v_i} \Rightarrow Sol(\mathbf{u_k}))) \vee ... \vee \\ ((\mathbf{U} = \mathbf{u_k} \Box\!\!\rightarrow \mathbf{V} = \mathbf{v_i}) \wedge (\mathbf{U} = \mathbf{u_k} \Box\!\!\rightarrow \mathbf{V} = \mathbf{v_i} \Rightarrow Sol(\mathbf{u_k})) \wedge (\mathbf{U} = \mathbf{u_1} \\ \Box\!\!\rightarrow \mathbf{V} = \mathbf{v_i} \Rightarrow Sol(\mathbf{u_1})) \wedge ... \wedge (\mathbf{U} = \mathbf{u_{k-1}} \Box\!\!\rightarrow \mathbf{V} = \mathbf{v_i} \Rightarrow Sol(\mathbf{u_{k-1}}))),\end{aligned} \tag{4}$$

where $R(\mathbf{U}) = \{\mathbf{u_1}, ..., \mathbf{u_k}\}$. As each disjunct of (4) can derive $Sol(\mathbf{u_1}) \vee ... \vee Sol(\mathbf{u_k})$, i.e., $\bigvee_{u\in R(U)}((\boldsymbol{U} = \boldsymbol{u} \wedge \boldsymbol{Y} = \boldsymbol{y} \wedge \boldsymbol{V}_{Y\cup\{X_1\}} = \boldsymbol{v}_{Y\cup\{X_1\}}) \Box\!\!\rightarrow X_1 = x_1 \wedge ... \wedge (\boldsymbol{U} = \boldsymbol{u} \wedge \boldsymbol{Y} = \boldsymbol{y} \wedge \boldsymbol{V}_{Y\cup\{X_m\}} = \boldsymbol{v}_{Y\cup\{X_m\}}) \Box\!\!\rightarrow X_m = x_m)$, the consequent can derive $Sol(\mathbf{u_1}) \vee ... \vee Sol(\mathbf{u_k})$. It follows that $\mathbf{V} = \mathbf{v_i} \Rightarrow \mathbf{Y} = \mathbf{y} \Box\!\!\rightarrow \mathbf{V} = \mathbf{v} \Rightarrow Sol(\mathbf{u_1}) \vee ... \vee Sol(\mathbf{u_k})$. Together with $\bigvee_{\mathbf{v_i}\in R(\mathbf{V})} \mathbf{V} = \mathbf{v_i}$, we have $\boldsymbol{Y} = \boldsymbol{y} \Box\!\!\rightarrow \boldsymbol{V} = \boldsymbol{v} \Rightarrow \bigvee_{u\in R(U)}((\boldsymbol{U} = \boldsymbol{u} \wedge \boldsymbol{Y} = \boldsymbol{y} \wedge \boldsymbol{V}_{Y\cup\{X_1\}} = \boldsymbol{v}_{Y\cup\{X_1\}}) \Box\!\!\rightarrow X_1 = x_1 \wedge ... \wedge (\boldsymbol{U} = \boldsymbol{u} \wedge \boldsymbol{Y} = \boldsymbol{y} \wedge \boldsymbol{V}_{Y\cup\{X_m\}} = \boldsymbol{v}_{Y\cup\{X_m\}}) \Box\!\!\rightarrow X_m = x_m)$ as a theorem. □

Proposition 1. *M^c is a Pearlian causal model.*

Proof. We show that F^c has a unique solution after any intervention $\boldsymbol{Y} = \boldsymbol{y}$ ($\boldsymbol{Y} \subseteq \mathbf{V}$) in any $\boldsymbol{u}$ by induction on $|\mathbf{V}| - |\boldsymbol{Y}|$:

$|\mathbf{V}| - |\boldsymbol{Y}| = 0$, suppose $\boldsymbol{Y} = \boldsymbol{y}$ is $\boldsymbol{V} = \boldsymbol{v}$, we have for any context $\boldsymbol{u}$, $(\boldsymbol{U} = \boldsymbol{u} \wedge \boldsymbol{V} = \boldsymbol{v}) \,\Box\!\!\rightarrow \boldsymbol{V} = \boldsymbol{v} \in \Gamma$ (due to EFFECTIVENESS and CONJUNCTION), then $F^c_{\boldsymbol{V}=\boldsymbol{v}}$ has a unique solution $\boldsymbol{v}$ given any context.

$|\mathbf{V}| - |\boldsymbol{Y}| = 1$, suppose $\boldsymbol{Y} = \boldsymbol{y}$ is $\boldsymbol{V}_X = \boldsymbol{v}_X$ for some $X \in \mathbf{V}$, given any context $\boldsymbol{u}$, we have $(\boldsymbol{U} = \boldsymbol{u} \wedge \boldsymbol{V}_X = \boldsymbol{v}_X) \,\Box\!\!\rightarrow X = x \in \Gamma$ for some $x \in R(X)$. Therefore $f_X(\boldsymbol{u}, \boldsymbol{v}_X) = x$ and $(\boldsymbol{U} = \boldsymbol{u} \wedge \boldsymbol{V}_X = \boldsymbol{v}_X) \,\Box\!\!\rightarrow (\boldsymbol{V}_X = \boldsymbol{v}_X \wedge X = x) \in \Gamma$, $F^c_{\boldsymbol{V}_X=\boldsymbol{v}_X}$ has solution $(\boldsymbol{v}_X, x)$ in $\boldsymbol{u}$, obviously it is unique.

Inductive Hypothesis: $|\mathbf{V}| - |\boldsymbol{Z}| = k$, given any context $\boldsymbol{u}$, according to $(\boldsymbol{U} = \boldsymbol{u} \wedge \boldsymbol{Z} = \boldsymbol{z}) \,\Box\!\!\rightarrow (\boldsymbol{Z} = \boldsymbol{z} \wedge \boldsymbol{V}_{\boldsymbol{Z}} = \boldsymbol{v}_{\boldsymbol{Z}}) \in \Gamma$, $F^c_{\boldsymbol{Z}=\boldsymbol{z}}$ has a unique solution $(\boldsymbol{z}, \boldsymbol{v}_{\boldsymbol{Z}})$ under any context $\boldsymbol{u}$.

$|\mathbf{V}| - |\boldsymbol{Y}| = k + 1 (k \geq 1)$, suppose for any context $\boldsymbol{u}$, $(\boldsymbol{U} = \boldsymbol{u} \wedge \boldsymbol{Y} = \boldsymbol{y}) \,\Box\!\!\rightarrow (\boldsymbol{Y} = \boldsymbol{y} \wedge \boldsymbol{V}_{\boldsymbol{Y}} = \boldsymbol{v}_{\boldsymbol{Y}}) \in \Gamma$, we prove $(\boldsymbol{y}, \boldsymbol{v}_{\boldsymbol{Y}})$ is a unique solution for $F^c_{\boldsymbol{Y}=\boldsymbol{y}}$ in $\boldsymbol{u}$. We can find $V_1 \in \boldsymbol{V}_{\boldsymbol{Y}}$ and suppose $V_1 = v_1$ is a conjunct of $\boldsymbol{V}_{\boldsymbol{Y}} = \boldsymbol{v}_{\boldsymbol{Y}}$. Then for $\boldsymbol{Y} = \boldsymbol{y}, V_1 = v_1$, due to COMPOSITION, $(\boldsymbol{U} = \boldsymbol{u} \wedge \boldsymbol{Y} = \boldsymbol{y} \wedge V_1 = v_1) \,\Box\!\!\rightarrow (\boldsymbol{Y} = \boldsymbol{y} \wedge \boldsymbol{V}_{\boldsymbol{Y} \cup \{V_1\}} = \boldsymbol{v}_{\boldsymbol{Y} \cup \{V_1\}}) \in \Gamma$. Then $(\boldsymbol{U} = \boldsymbol{u} \wedge \boldsymbol{Y} = \boldsymbol{y} \wedge V_1 = v_1) \,\Box\!\!\rightarrow (\boldsymbol{Y} = \boldsymbol{y} \wedge V_1 = v_1 \wedge \boldsymbol{V}_{\boldsymbol{Y} \cup \{V_1\}} = \boldsymbol{v}_{\boldsymbol{Y} \cup \{V_1\}}) \in \Gamma$. According to Inductive Hypothesis, $(\boldsymbol{y}, v_1, \boldsymbol{v}_{\boldsymbol{Y} \cup \{V_1\}})$ is a unique solution for $F^c_{\boldsymbol{Y}=\boldsymbol{y}, V_1=v_1}$ in $\boldsymbol{u}$. That means for any $Z \in \boldsymbol{V}_{\boldsymbol{Y} \cup \{V_1\}}$, f_Z can be fulfilled by $(\boldsymbol{y}, v_1, \boldsymbol{v}_{\boldsymbol{Y} \cup \{V_1\}})$ or $(\boldsymbol{y}, \boldsymbol{v}_{\boldsymbol{Y}})$ in $\boldsymbol{u}$. We only need to check whether f_{V_1} can be fulfilled by $(\boldsymbol{y}, v_1, \boldsymbol{v}_{\boldsymbol{Y} \cup \{V_1\}})$. If so, we can have for any $W \in \boldsymbol{V}_{\boldsymbol{Y}}$, f_W can be satisfied by $(\boldsymbol{y}, v_1, \boldsymbol{v}_{\boldsymbol{Y} \cup \{V_1\}})$. Therefore $(\boldsymbol{y}, \boldsymbol{v}_{\boldsymbol{Y}})$ is the solution for $F^c_{\boldsymbol{Y}=\boldsymbol{y}}$ in $\boldsymbol{u}$. To obtain this result, we can find another $V_2 \in \boldsymbol{V}_{\boldsymbol{Y}}$ and follow the same argument for V_1. Then $(\boldsymbol{y}, v_2, \boldsymbol{v}_{\boldsymbol{Y} \cup \{V_2\}})$ is a unique solution for $F^c_{\boldsymbol{Y}=\boldsymbol{y}, V_2=v_2}$ in $\boldsymbol{u}$ ($V_2 = v_2$ is a conjunct of $\boldsymbol{V}_{\boldsymbol{Y}} = \boldsymbol{v}_{\boldsymbol{Y}}$). We can know that f_{V_1} can be fulfilled by $(\boldsymbol{y}, v_2, \boldsymbol{v}_{\boldsymbol{Y} \cup \{V_1\}})$ or $(\boldsymbol{y}, \boldsymbol{v}_{\boldsymbol{Y}})$.

Then we need to prove that $(\boldsymbol{y}, \boldsymbol{v}_{\boldsymbol{Y}})$ is unique for $F^c_{\boldsymbol{Y}=\boldsymbol{y}}$ in $\boldsymbol{u}$. Suppose there is another solution $(\boldsymbol{y}, \boldsymbol{v}'_{\boldsymbol{Y}})$ for $F^c_{\boldsymbol{Y}=\boldsymbol{y}}$ in $\boldsymbol{u}$. That means $(\boldsymbol{U} = \boldsymbol{u} \wedge \boldsymbol{Y} = \boldsymbol{y} \wedge \boldsymbol{V}_{\boldsymbol{Y} \cup \{X_1\}} = \boldsymbol{v}'_{\boldsymbol{Y} \cup \{X_1\}}) \,\Box\!\!\rightarrow X_1 = x_1 \wedge ... \wedge (\boldsymbol{U} = \boldsymbol{u} \wedge \boldsymbol{Y} = \boldsymbol{y} \wedge \boldsymbol{V}_{\boldsymbol{Y} \cup \{X_m\}} = \boldsymbol{v}'_{\boldsymbol{Y} \cup \{X_m\}}) \,\Box\!\!\rightarrow X_m = x_m \in \Gamma$ ($\boldsymbol{V}_{\boldsymbol{Y}} = \{X_1, ..., X_m\}, X_1 = x_1 \wedge ... \wedge X_m = x_m$ is $\boldsymbol{v}'_{\boldsymbol{Y}}$). According to SOLUTION*, $(\boldsymbol{U} = \boldsymbol{u} \wedge \boldsymbol{Y} = \boldsymbol{y}) \,\Box\!\!\rightarrow (\boldsymbol{Y} = \boldsymbol{y} \wedge \boldsymbol{V}_{\boldsymbol{Y}} = \boldsymbol{v}'_{\boldsymbol{Y}}) \in \Gamma$. But we already have $(\boldsymbol{U} = \boldsymbol{u} \wedge \boldsymbol{Y} = \boldsymbol{y}) \,\Box\!\!\rightarrow (\boldsymbol{Y} = \boldsymbol{y} \wedge \boldsymbol{V}_{\boldsymbol{Y}} = \boldsymbol{v}_{\boldsymbol{Y}}) \in \Gamma$. There must be some variable V_i such that $(\boldsymbol{U} = \boldsymbol{u} \wedge \boldsymbol{Y} = \boldsymbol{y}) \,\Box\!\!\rightarrow V_i = v_i \in \Gamma$ (due to CONJUNCTION, $V_i = v_i$ is a conjunct of $\boldsymbol{V}_{\boldsymbol{Y}} = \boldsymbol{v}_{\boldsymbol{Y}}$) and $(\boldsymbol{U} = \boldsymbol{u} \wedge \boldsymbol{Y} = \boldsymbol{y}) \,\Box\!\!\rightarrow V_i = v'_i \in \Gamma$ ($v'_i \neq v_i, V_i = v'_i$ is a conjunct of $\boldsymbol{V}_{\boldsymbol{Y}} = \boldsymbol{v}'_{\boldsymbol{Y}}$). This result contradicts with EQUALITY. □

Theorem 2 (Truth Lemma). *For any $\psi \in \mathcal{L}(S)$, $\psi \in \Gamma$ iff $\langle M^c, P^c, \boldsymbol{u}^c \rangle \models \psi$.*

Proof. We prove this lemma by induction on ψ. Consider the case that ψ is $\boldsymbol{Y} = \boldsymbol{y} \,\Box\!\!\rightarrow \psi'$ for some $\boldsymbol{Y} \subseteq \mathbf{U} \cup \mathbf{V}$ and ψ'. The spirit of the proof is to rewrite ψ' as a conjunctive normal form ψ'', and then reduce $\boldsymbol{Y} = \boldsymbol{y} \,\Box\!\!\rightarrow \psi''$ to several conjuncts of the form $\boldsymbol{Y} = \boldsymbol{y} \,\Box\!\!\rightarrow \boldsymbol{Z}_\vee = \boldsymbol{z}$ ($\boldsymbol{Z}_\vee = \boldsymbol{z}$ represents the formulas like $Z_0 = z_0 \vee ... \vee Z_{n'} = z_{n'}$), we prove $\boldsymbol{Y} = \boldsymbol{y} \,\Box\!\!\rightarrow \boldsymbol{Z}_\vee = \boldsymbol{z} \in \Gamma \iff \langle M^c, P^c, \boldsymbol{u}^c \rangle \models \boldsymbol{Y} = \boldsymbol{y} \,\Box\!\!\rightarrow \boldsymbol{Z}_\vee = \boldsymbol{z}$.

According to RE, $\boldsymbol{Y}=\boldsymbol{y} \,\Box\!\!\rightarrow \psi' \Leftrightarrow \boldsymbol{Y}=\boldsymbol{y} \,\Box\!\!\rightarrow \psi''$ is a **SPCL(S)** theorem, we need to show that $\boldsymbol{Y}=\boldsymbol{y} \,\Box\!\!\rightarrow \psi'' \in \Gamma \iff \langle M^c, P^c, \boldsymbol{u}^c\rangle \vDash \boldsymbol{Y}=\boldsymbol{y} \,\Box\!\!\rightarrow \psi''$. We have CONJUNCTION, so $\boldsymbol{Y}=\boldsymbol{y} \,\Box\!\!\rightarrow \psi''$ can be separated into a conjunction of formulas of the form $\boldsymbol{Y}=\boldsymbol{y} \,\Box\!\!\rightarrow \psi_i$ where ψ_i is a disjunction of formulas of the form $Z=z$ and their negations. According to DEFINITENESS and EQUALITY, we have theorem $\neg X=x \Leftrightarrow X=x_1 \vee \ldots \vee X=x_k$ where $x \neq x_i$, $i=1,\ldots,k$ and $R(X)=\{x,x_1,\ldots,x_k\}$, then $\varphi' \vee \neg X=x \Leftrightarrow \varphi' \vee (X=x_1 \vee \ldots \vee X=x_k)$ is a theorem. Due to RE, we have theorem $\boldsymbol{Y}=\boldsymbol{y} \,\Box\!\!\rightarrow (\varphi' \vee \neg X=x) \Leftrightarrow \boldsymbol{Y}=\boldsymbol{y} \,\Box\!\!\rightarrow (\varphi' \vee (X=x_1 \vee \ldots \vee X=x_k))$. At last we only need to prove $\boldsymbol{Y}=\boldsymbol{y} \,\Box\!\!\rightarrow \boldsymbol{Z}_\vee=\boldsymbol{z} \in \Gamma \iff \langle M^c, P^c, \boldsymbol{u}^c\rangle \vDash \boldsymbol{Y}=\boldsymbol{y} \,\Box\!\!\rightarrow \boldsymbol{Z}_\vee=\boldsymbol{z}$.

"$\Rightarrow$" Suppose that $\boldsymbol{Y}=\boldsymbol{y} \,\Box\!\!\rightarrow \boldsymbol{Z}_\vee=\boldsymbol{z} \in \Gamma$, we need to prove that after updating $\boldsymbol{v}^c$ (by SOLUTION*, $\boldsymbol{u}^c$ can derive $\boldsymbol{v}^c$ in M^c), under any context $\boldsymbol{u}$ that is consistent with $\boldsymbol{v}^c$, $\boldsymbol{Y}=\boldsymbol{y} \,\Box\!\!\rightarrow \boldsymbol{Z}_\vee=\boldsymbol{z}$ holds. At first we need to find out those contexts that are consistent with $\boldsymbol{v}^c$. According to POSSIBILITY, as $\boldsymbol{v}^c \in \Gamma$, there are contexts $\boldsymbol{u}$ such that $(\boldsymbol{u} \wedge \boldsymbol{v}^c_{\bar{X}_1}) \,\Box\!\!\rightarrow X_1=x_1 \wedge \ldots \wedge (\boldsymbol{u} \wedge \boldsymbol{v}^c_{\bar{X}_n}) \,\Box\!\!\rightarrow X_n=x_n \in \Gamma$ ($\mathbf{V}=\{X_1,\ldots,X_n\}$). Then we gather those $\boldsymbol{u}$ to be a set T. For each that $\boldsymbol{u}^* \in T$, according to the definition of structural equations of the canonical model, we know that in that model under $\boldsymbol{u}^*$, the equations have a solution $\boldsymbol{v}^c$. That is, $\boldsymbol{u}^*$ is consistent with $\boldsymbol{v}^c$.

The fact that $\boldsymbol{u}^*$ is consistent with $\boldsymbol{v}^c$ tells us that $\langle M^c, P^c, \boldsymbol{u}^c\rangle \vDash \boldsymbol{u}^* \,\Box\!\!\rightarrow \boldsymbol{v}^c$. According to the soundness of SOLUTION*, we then have

$$\langle M^c, P^c, \boldsymbol{u}^c\rangle \vDash (\boldsymbol{u}^* \wedge \boldsymbol{v}^c_{\bar{X}_1}) \,\Box\!\!\rightarrow X_1=x_1 \wedge \ldots \wedge (\boldsymbol{u}^* \wedge \boldsymbol{v}^c_{\bar{X}_n}) \,\Box\!\!\rightarrow X_n=x_n.$$

That is to say, the equations in the model are consistent after we input $\boldsymbol{u}^*$ and $\boldsymbol{v}^c$ into them. Therefore, $(\boldsymbol{u}^* \wedge \boldsymbol{v}^c_{\bar{X}_1}) \,\Box\!\!\rightarrow X_1=x_1 \wedge \ldots \wedge (\boldsymbol{u}^* \wedge \boldsymbol{v}^c_{\bar{X}_n}) \,\Box\!\!\rightarrow X_n=x_n \in \Gamma$. Hence, $\boldsymbol{u}^* \,\Box\!\!\rightarrow \boldsymbol{v}^c \in \Gamma$. According to TRANSFORMATION and the fact that $\boldsymbol{v}^c \in \Gamma$,

$$\boldsymbol{Y}=\boldsymbol{y} \,\Box\!\!\rightarrow \boldsymbol{Z}_\vee=\boldsymbol{z} \Leftrightarrow \bigwedge_{\boldsymbol{u} \in R(\boldsymbol{U})} (\boldsymbol{U}=\boldsymbol{u} \,\Box\!\!\rightarrow \boldsymbol{V}=\boldsymbol{v}^c \Rightarrow$$
$$(\boldsymbol{U}=\boldsymbol{u} \wedge \boldsymbol{Y}=\boldsymbol{y}) \,\Box\!\!\rightarrow \boldsymbol{Z}_\vee=\boldsymbol{z}) \in \Gamma,$$

so $\bigwedge_{\boldsymbol{u} \in R(\boldsymbol{U})} (\boldsymbol{U}=\boldsymbol{u} \,\Box\!\!\rightarrow \boldsymbol{V}=\boldsymbol{v}^c \Rightarrow (\boldsymbol{U}=\boldsymbol{u} \wedge \boldsymbol{Y}=\boldsymbol{y}) \,\Box\!\!\rightarrow \boldsymbol{Z}_\vee=\boldsymbol{z}) \in \Gamma$, $(\boldsymbol{u}^* \wedge \boldsymbol{y}) \,\Box\!\!\rightarrow \boldsymbol{Z}_\vee=\boldsymbol{z} \in \Gamma$. Due to DISJUNCTION, $\bigvee_{Z_i \in \boldsymbol{Z}} (\boldsymbol{u}^* \wedge \boldsymbol{y}) \,\Box\!\!\rightarrow Z_i=z_i \in \Gamma$ (z_i is the corresponding value of Z_i in $\boldsymbol{Z}_\vee=\boldsymbol{z}$). We suppose that $(\boldsymbol{u}^* \wedge \boldsymbol{y}) \,\Box\!\!\rightarrow Z_0=z_0 \in \Gamma$. According to FUNCTIONALITY, for each $X_i \in \mathbf{V}$, $\bigvee_{x_i' \in R(X_i)} (\boldsymbol{u}^* \wedge \boldsymbol{y}) \,\Box\!\!\rightarrow X_i=x_i'$, then there is some $\boldsymbol{v} \in R(\boldsymbol{V})$ s.t. $(\boldsymbol{u}^* \wedge \boldsymbol{y}) \,\Box\!\!\rightarrow \boldsymbol{v} \in \Gamma$. Because of EQUALITY, Z_0 takes on z_0 in $\boldsymbol{v}$. Due to SOLUTION*, we have

$$(\boldsymbol{u}^* \wedge \boldsymbol{y} \wedge \boldsymbol{v}_{\boldsymbol{Y} \cup \{X_1\}}) \,\Box\!\!\rightarrow X_1=x_1' \wedge \ldots \wedge (\boldsymbol{u}^* \wedge \boldsymbol{y} \wedge \boldsymbol{v}_{\boldsymbol{Y} \cup \{X_m\}}) \,\Box\!\!\rightarrow X_m=x_m' \in \Gamma$$

($m=|\mathbf{V}|-|\boldsymbol{Y}|$). That gives us the corresponding assignment for the equations in the canonical model, so we can get $\langle M^c, P^c, \boldsymbol{u}^c\rangle \vDash (\boldsymbol{u}^* \wedge \boldsymbol{y}) \,\Box\!\!\rightarrow \boldsymbol{v}$. As $Z_0=z_0$ is a conjunct of $\boldsymbol{v}$, so $\langle M^c, P^c, \boldsymbol{u}^c\rangle \vDash (\boldsymbol{u}^* \wedge \boldsymbol{y}) \,\Box\!\!\rightarrow Z_0=z_0$ (due to CONJUNCTION), $\langle M^c, P^c, \boldsymbol{u}^c\rangle \vDash (\boldsymbol{u}^* \wedge \boldsymbol{y}) \,\Box\!\!\rightarrow \boldsymbol{Z}_\vee=\boldsymbol{z}$ (due to DISJUNCTION). For any context

$\boldsymbol{u}$ such that $\langle M^c, P^c, \boldsymbol{u}^c\rangle \models \boldsymbol{u} \,\Box\!\!\rightarrow \boldsymbol{v}^c$, $\boldsymbol{u}$ is consistent with $\boldsymbol{v}^c$, $\boldsymbol{u}$ is some $\boldsymbol{u}^* \in T$, so $\langle M^c, P^c, \boldsymbol{u}^c\rangle \models (\boldsymbol{u} \wedge \boldsymbol{y}) \,\Box\!\!\rightarrow \boldsymbol{Z}_\vee = \boldsymbol{z}$. Therefore we have

$$\langle M^c, P^c, \boldsymbol{u}^c\rangle \models \bigwedge_{\boldsymbol{u}\in R(\boldsymbol{U})} (\boldsymbol{U} = \boldsymbol{u} \,\Box\!\!\rightarrow \boldsymbol{V} = \boldsymbol{v}^c \Rightarrow (\boldsymbol{U} = \boldsymbol{u} \wedge \boldsymbol{Y} = \boldsymbol{y}) \,\Box\!\!\rightarrow \boldsymbol{Z}_\vee = \boldsymbol{z}).$$

According to TRANSFORMATION and the fact that $\langle M^c, P^c, \boldsymbol{u}^c\rangle \models \boldsymbol{V} = \boldsymbol{v}^c$, $\langle M^c, P^c, \boldsymbol{u}^c\rangle \models \boldsymbol{Y} = \boldsymbol{y} \,\Box\!\!\rightarrow \boldsymbol{Z}_\vee = \boldsymbol{z}$.

"$\Leftarrow$" Suppose $\langle M^c, P^c, \boldsymbol{u}^c\rangle \models \boldsymbol{Y} = \boldsymbol{y} \,\Box\!\!\rightarrow \boldsymbol{Z}_\vee = \boldsymbol{z}$, we need to show that $\bigwedge_{\boldsymbol{u}\in R(\boldsymbol{U})}(\boldsymbol{U} = \boldsymbol{u} \,\Box\!\!\rightarrow \boldsymbol{V} = \boldsymbol{v}^c \Rightarrow (\boldsymbol{U} = \boldsymbol{u} \wedge \boldsymbol{Y} = \boldsymbol{y}) \,\Box\!\!\rightarrow \boldsymbol{Z}_\vee = \boldsymbol{z}) \in \Gamma$. For any context $\boldsymbol{u}^*$ such that $\boldsymbol{u}^* \,\Box\!\!\rightarrow \boldsymbol{v}^c \in \Gamma$, we have $(\boldsymbol{u}^* \wedge \boldsymbol{v}^c_{\overline{X_1}}) \,\Box\!\!\rightarrow X_1 = x_1 \wedge ... \wedge (\boldsymbol{u}^* \wedge \boldsymbol{v}^c_{\overline{X_n}}) \,\Box\!\!\rightarrow X_n = x_n \in \Gamma$ by using SOLUTION*. According to the definition of the equations in the canonical model, $\langle M^c, P^c, \boldsymbol{u}^c\rangle \models \boldsymbol{u}^* \,\Box\!\!\rightarrow \boldsymbol{v}^c$.

Due to the soundness of TRANSFORMATION and the fact that $\langle M^c, P^c, \boldsymbol{u}^c\rangle \models \boldsymbol{V} = \boldsymbol{v}^c$,

$$\langle M^c, P^c, \boldsymbol{u}^c\rangle \models \boldsymbol{Y} = \boldsymbol{y} \,\Box\!\!\rightarrow \boldsymbol{Z}_\vee = \boldsymbol{z} \Leftrightarrow$$
$$\bigwedge_{\boldsymbol{u}\in R(\boldsymbol{U})} (\boldsymbol{U} = \boldsymbol{u} \,\Box\!\!\rightarrow \boldsymbol{V} = \boldsymbol{v}^c \Rightarrow (\boldsymbol{U} = \boldsymbol{u} \wedge \boldsymbol{Y} = \boldsymbol{y}) \,\Box\!\!\rightarrow \boldsymbol{Z}_\vee = \boldsymbol{z}).$$

As $\langle M^c, P^c, \boldsymbol{u}^c\rangle \models \boldsymbol{Y} = \boldsymbol{y} \,\Box\!\!\rightarrow \boldsymbol{Z}_\vee = \boldsymbol{z}$, we obtain $\langle M^c, P^c, \boldsymbol{u}^c\rangle \models \bigwedge_{\boldsymbol{u}\in R(\boldsymbol{U})}(\boldsymbol{U} = \boldsymbol{u} \,\Box\!\!\rightarrow \boldsymbol{V} = \boldsymbol{v}^c \Rightarrow (\boldsymbol{U} = \boldsymbol{u} \wedge \boldsymbol{Y} = \boldsymbol{y}) \,\Box\!\!\rightarrow \boldsymbol{Z}_\vee = \boldsymbol{z})$. That is, for any context $\boldsymbol{u}$ such that $\langle M^c, P^c, \boldsymbol{u}^c\rangle \models \boldsymbol{u} \,\Box\!\!\rightarrow \boldsymbol{v}^c$, $\langle M^c, P^c, \boldsymbol{u}^c\rangle \models (\boldsymbol{U} = \boldsymbol{u} \wedge \boldsymbol{Y} = \boldsymbol{y}) \,\Box\!\!\rightarrow \boldsymbol{Z}_\vee = \boldsymbol{z}$. Therefore $\langle M^c, P^c, \boldsymbol{u}^c\rangle \models (\boldsymbol{U} = \boldsymbol{u}^* \wedge \boldsymbol{Y} = \boldsymbol{y}) \,\Box\!\!\rightarrow \boldsymbol{Z}_\vee = \boldsymbol{z}$, there is some disjunct $Z_i = z_i$ in $\boldsymbol{Z}_\vee = \boldsymbol{z}$ such that $\langle M^c, P^c, \boldsymbol{u}^c\rangle \models (\boldsymbol{U} = \boldsymbol{u}^* \wedge \boldsymbol{Y} = \boldsymbol{y}) \,\Box\!\!\rightarrow Z_i = z_i$ (due to the soundness of DISJUNCTION). Suppose $\langle M^c, P^c, \boldsymbol{u}^c\rangle \models (\boldsymbol{U} = \boldsymbol{u}^* \wedge \boldsymbol{Y} = \boldsymbol{y}) \,\Box\!\!\rightarrow Z_0 = z_0$, as there exists $\boldsymbol{v}$ such that $\langle M^c, P^c, \boldsymbol{u}^c\rangle \models (\boldsymbol{U} = \boldsymbol{u}^* \wedge \boldsymbol{Y} = \boldsymbol{y}) \,\Box\!\!\rightarrow \boldsymbol{v}$, $Z_0 = z_0$ is a conjunct of $\boldsymbol{v}$. Due to the soundness of SOLUTION*,

$$\langle M^c, P^c, \boldsymbol{u}^c\rangle \models (\boldsymbol{u}^* \wedge \boldsymbol{y} \wedge \boldsymbol{v}_{\overline{\boldsymbol{Y}\cup\{X_1\}}}) \,\Box\!\!\rightarrow X_1 = x_1 \wedge ... \wedge$$
$$(\boldsymbol{u}^* \wedge \boldsymbol{y} \wedge \boldsymbol{v}_{\overline{\boldsymbol{Y}\cup\{X_m\}}}) \,\Box\!\!\rightarrow X_m = x_m.$$

That tells us the corresponding assignment of the equations in the canonical model. So

$$(\boldsymbol{U} = \boldsymbol{u}^* \wedge \boldsymbol{Y} = \boldsymbol{y} \wedge \boldsymbol{V}_{\overline{\boldsymbol{Y}\cup\{X_1\}}} = \boldsymbol{v}_{\overline{\boldsymbol{Y}\cup\{X_1\}}}) \,\Box\!\!\rightarrow X_1 = x_1 \wedge ... \wedge$$
$$(\boldsymbol{U} = \boldsymbol{u}^* \wedge \boldsymbol{Y} = \boldsymbol{y} \wedge \boldsymbol{V}_{\overline{\boldsymbol{Y}\cup\{X_m\}}} = \boldsymbol{v}_{\overline{\boldsymbol{Y}\cup\{X_m\}}}) \,\Box\!\!\rightarrow X_m = x_m \in \Gamma,$$

we can have $(\boldsymbol{u}^* \wedge \boldsymbol{y}) \,\Box\!\!\rightarrow \boldsymbol{v} \in \Gamma$. Due to CONJUNCTION, $(\boldsymbol{U} = \boldsymbol{u}^* \wedge \boldsymbol{Y} = \boldsymbol{y}) \,\Box\!\!\rightarrow Z_0 = z_0 \in \Gamma$. Then $(\boldsymbol{U} = \boldsymbol{u}^* \wedge \boldsymbol{Y} = \boldsymbol{y}) \,\Box\!\!\rightarrow Z_0 = z_0 \vee (\boldsymbol{U} = \boldsymbol{u}^* \wedge \boldsymbol{Y} = \boldsymbol{y}) \,\Box\!\!\rightarrow \boldsymbol{Z}_{0\vee} = \boldsymbol{z}_0 \in \Gamma$ ($\boldsymbol{Z}_0 = \boldsymbol{Z} - \{Z_0\}$, $\boldsymbol{z}_0$ is the value of $\boldsymbol{Z}_0$ in $\boldsymbol{z}$), $(\boldsymbol{U} = \boldsymbol{u}^* \wedge \boldsymbol{Y} = \boldsymbol{y}) \,\Box\!\!\rightarrow \boldsymbol{Z}_\vee = \boldsymbol{z} \in \Gamma$ according to DISJUNCTION. Then we have proved that

$$\bigwedge_{\boldsymbol{u}\in R(\boldsymbol{U})} (\boldsymbol{U} = \boldsymbol{u} \,\Box\!\!\rightarrow \boldsymbol{V} = \boldsymbol{v}^c \Rightarrow (\boldsymbol{U} = \boldsymbol{u} \wedge \boldsymbol{Y} = \boldsymbol{y}) \,\Box\!\!\rightarrow \boldsymbol{Z}_\vee = \boldsymbol{z}) \in \Gamma,$$

due to TRANSFORMATION and the fact that $\boldsymbol{v}^c \in \Gamma$, we have $\boldsymbol{Y} = \boldsymbol{y} \Box\!\!\rightarrow Z_{\vee} = \boldsymbol{z} \in \Gamma$.

Inductive Hypothesis: For any $\varphi^* \in \mathcal{L}(S)$, $\varphi^* \in \Gamma \iff \langle M^c, P^c, \boldsymbol{u}^c \rangle \vDash \varphi^*$.

If ψ has the form of $\neg\psi_0$ or $\psi_1 \wedge \psi_2$, the proof is trivial according to the properties of maximal consistent sets. □

References

1. Barbero, F., Virtema, J.: Strongly complete axiomatization for a logic with probabilistic interventionist counterfactuals (2023). https://arxiv.org/abs/2304.02964
2. Beckers, S.: Causal sufficiency and actual causation. J. Philos. Log., 1–34 (2021). https://doi.org/10.1007/s10992-021-09601-z
3. Beckers, S.: Nondeterministic causal models (2024). https://arxiv.org/abs/2405.14001
4. Bennett, J.: A Philosophical Guide to Conditionals. Clarendon Press, Oxford, UK (2003)
5. Blackburn, S.: How can we tell whether a commitment has a truth condition? In: Travis, C. (ed.) Meaning and Interpretation, pp. 201–232. Basil Blackwell, Oxford, UK (1986)
6. Cartwright, N.: Causal laws and effective strategies. Noûs **13**(4), 419–437 (1979). http://www.jstor.org/stable/2215337
7. Eells, E.: Probabilistic Causality. Cambridge Studies in Probability, Induction and Decision Theory, Cambridge University Press (1991). https://doi.org/10.1017/CBO9780511570667
8. Fenton-Glynn, L.: A probabilistic analysis of causation. Br. J. Philos. Sci. **62**(2), 343–392 (2011)
9. Fenton-Glynn, L.: A proposed probabilistic extension of the Halpern and Pearl definition of actual cause. Br. J. Philos. Sci. **68**(4), 1061–1124 (2017)
10. Galliani, P.: Dependence Logic. In: Zalta, E.N., Nodelman, U. (eds.) The Stanford Encyclopedia of Philosophy. Metaphysics Research Lab, Stanford University, Summer 2024 edn. (2024). https://plato.stanford.edu/archives/sum2024/entries/logic-dependence/
11. Glymour, C., Wimberly, F.: Actual causes and thought experiments. In: Campbell, J.K., O'Rourke, M., Silverstein, H.S. (eds.) Causation and Explanation. MIT Press (2007)
12. Hájek, A.: What conditional probability could not be. Synthese **137**, 273–323 (2004)
13. Hall, N.: Structural equations and causation. Philos. Stud. **132**(1), 109–136 (2007)
14. Halpern, J.Y.: Axiomatizing causal reasoning. J. Artif. Intell. Res. **12**, 317–337 (2000)
15. Halpern, J.Y.: A modification of the Halpern-Pearl definition of causality. In: Proceedings of the 24th International Conference on Artificial Intelligence, pp. 3022–3033. IJCAI 2015, AAAI Press (2015)
16. Halpern, J.Y.: Actual Causality. MIT Press, Cambridge, MA (2016)
17. Halpern, J.Y., Pearl, J.: Causes and explanations: a structural-model approach. part I: causes. Proceedings Seventeenth Conference on Uncertainty in Artificial Intelligence (UAI 2001), pp. 194–202 (2001)
18. Halpern, J.Y., Pearl, J.: Causes and explanations: a structural-model approach. part I: causes. Br. J. Philos. Sci. **56**(4), 843–887 (2005). https://doi.org/10.1093/bjps/axi147

19. Hawthorne, J.: Nonmonotonic conditionals that behave like conditional probabilities above a threshold. J. Appl. Logic **5**(4), 625–637 (2007). selected papers from the 4th International Workshop on Computational Models of Scientific Reasoning and Applications
20. Hawthorne, J.,C.D., Makinson, D.: The quantitative/qualitative watershed for rules of uncertain inference. Studia Logica **86** (2007). https://doi.org/10.1007/s11225-007-9061-x
21. Hitchcock, C.: The intransitivity of causation revealed in equations and graphs. J. Philos. **98**(6), 273–299 (2001). http://www.jstor.org/stable/2678432
22. Hitchcock, C.: Prevention, preemption, and the principle of sufficient reason. Philos. Rev. **116**(4), 495–532 (2007). https://doi.org/10.1215/00318108-2007-012
23. Ibeling, D., Icard, T.: Probabilistic reasoning across the causal hierarchy. CoRR abs/2001.02889 (2020). http://arxiv.org/abs/2001.02889
24. Kvart, I.: The causal-process-chance-based analysis of counterfactuals. Unpublished manuscript
25. Kvart, I.: Causation: Probabilistic and Counterfactual Analyses. Collins, Hall, and Paul, pp. 359–387 (2004)
26. Leitgeb, H.: A probabilistic semantics for counterfactuals. part A. Rev. Symbolic Logic **5**, 26–84 (2012)
27. Lewis, D.: Postscripts to 'causation'. In: Philosophical Papers Volume II, pp. 172–213. Oxford University Press (1986)
28. Paris, J.B., Simmonds, R.: Review of Symbolic logic, vol. 2 (2009)
29. Pearl, J.: Causality: Models, Reasoning, and Inference. Cambridge University Press, New York (2009)
30. Reichenbach, H.: The Direction of Time. Dover Publications (1956)
31. Steel, D.: Indeterminism and the causal markov condition. Br. J. Philos. Sci. **56**(1), 3–26 (2005)
32. Weslake, B.: A partial theory of actual causation. Br. J. Philos. Sci. **66**(3), 521–557 (2015)
33. Woodward, J.: Making Things Happen: A Theory of Causal Explanation. Oxford University Press, Oxford, U.K. (2003)

Disjunction, Juxtaposition, and Alternative Questions in Mandarin

Rong He(✉)

Tsinghua University, Beijing, China
her22@mails.tsinghua.edu.cn

Abstract. It is widely assumed that Mandarin lexically distinguishes between the disjunctions used in alternative questions (*háishì*) and disjunctive propositions (*huòzhě*). This paper challenges this standard view by presenting a decompositional analysis of *háishì*. Specifically, we argue that the *hái* and *shì* in *háishì* are both semantically active, with *hái* conveying additivity and *shì* exhaustivity. The nesting of these two particles gives rise to the mutual exclusivity presupposition, a presupposition observed cross-linguistically in alternative questions. As a result, *háishì* should not be regarded as an interrogative disjunctor but rather as an adverb taking scope over a full CP. This analysis offers several empirical and theoretical advantages: (i) it resolves the distributional contrast between *háishì* and *huòzhě* without resorting to ad hoc assumptions; (ii) it provides a unified syntactic strategy for constructing Mandarin alternative questions; and (iii) it extends to the two additional non-disjunctive uses of *háishì*.

Keywords: Disjunction · Juxtaposition · Alternative Question · Mandarin

1 Introduction

Many languages differentiate between interrogative and standard disjunction ([17, 33]), as observed in languages like Finnish, Polish, and Vietnamese. Mandarin seems to fit this typological pattern as well. As illustrated in (1), *háishì* is exclusively employed in alternative questions, whereas *huòzhě* is utilized in disjunctive polar questions and declarative sentences.

(1) a. Zhāng Sān xǐhuān kāfēi háishì chá
Zhang San like coffee HAISHI tea
'Does Zhang San like coffee or tea?'
(AltQ:✓; PolQ: *; Decl: *)

b. Zhāng Sān xǐhuān kāfēi huòzhě chá
Zhang San like coffee HUOZHE tea
'Does Zhang San like coffee or tea? /Zhang San likes coffee or tea.'
(AltQ: *; PolQ:✓; Decl:✓)

J. Yan et al. (Eds.): TLLM 2024, LNCS 15410, pp. 51–68, 2025.
https://doi.org/10.1007/978-3-031-86054-6_3

This distributional contrast has been extensively explored in the literature, with both syntactic and semantic accounts proposed. From a syntactic standpoint, it is argued that *háishì* bears a [+*wh*] or [+*Q*] feature that must be checked by a question complementizer ([21, 24, 51], among others). Grounded in the two-dimensional Roothian Alternative Semantics framework ([46, 47]), [15] proposes that both disjunctors project their disjuncts as a set of alternatives, with no defined ordinary value. The alternatives projected by *háishì* are interpreted by a question operator, whereas those projected by *huòzhě* are interpreted by an existential operator, which is enforced by syntactic feature-checking.

This paper proposes a semantic solution through a decompositional analysis of *háishì*, positing that it morphologically consists of two particles: *hái* and *shì*. Specifically, *hái* conveys additivity, analogous to the German *noch* ([2]), while *shì* encodes exhaustivity, introducing semantics akin to English *it*-clefts ([14, 55]). The nesting of these two particles gives rise to the mutual exclusivity presupposition observed cross-linguistically in alternative questions. This presupposition, together with the proposed semantics of *huòzhě*, effectively addresses the distributional contrast without relying on previous assumptions regarding syntax and lexical semantics.

Additionally, this decompositional analysis offers several empirical and theoretical advantages that set it apart from previous accounts. First, it allows for a unified syntactic strategy for constructing Mandarin alternative questions. Mandarin reference grammar suggests two strategies: *háishì* disjunction and juxtaposition ([24, p. 243]). For instance, as shown in (2b), the juxtaposition of two VPs can generate alternative questions. By decomposing *háishì*, we propose that φ *háishì* ψ is derived from the juxtaposition of φ and *háishì* ψ, thereby establishing that all Mandarin alternative questions are uniformly derived through juxtaposition.

(2) a. Zhāng Sān xǐhuān kāfēi háishì chá?
Zhang San like coffee HAISHI tea
b. Zhāng Sān xǐhuān kāfēi xǐhuān chá?
Zhang San like coffee like tea
'Does Zhang San like coffee or tea?'

Second, this analysis extends to the two non-disjunctive uses of *háishì*: continuative and suggestive, as exemplified in (3). To the best of my knowledge, this paper is the first to propose a uniform analysis encompassing all three uses of *háishì*.

(3) a. Zhāng Sān xǐhuān kāfēi háishì xǐhuān chá? [disjunctive]
Zhang San like coffee HAISHI like tea
'Does Zhang San like coffee or tea?'

b. Zhāng Sān háishì zài túshūguǎn. [continuative]
Zhang San HAISHI in library
'Zhang San is still in the library'

c. Wǒmen háishì qù Běijīng ba! [suggestive]
we HAISHI go Beijing SFP
'Let's go to Beijing instead!'

This paper is structured as follows: Sect. 2 outlines our decompositional analysis of *háishì* and the semantic composition of alternative questions with *háishì*. Section 3 presents how our approach contributes to the existing discussion on the semantics and syntax of alternative questions, particularly addressing the distributional contrast and demonstrating that Mandarin alternative questions are uniformly derived through juxtaposition. Section 4 extends the decompositional analysis to non-disjunctive uses of *háishì*. Finally, Sect. 5 concludes the paper.

2 Decompositional Analysis

This section elaborates on the decompositional analysis of *háishì* as outlined earlier.

2.1 Additive *Hái*

Like its German counterpart *noch*, *hái* has a variety of uses, including continuative, marginal, scalar, and additive functions, as illustrated in (4). We propose that the *hái* in *háishì* is the additive *hái*, as indicated by the analogous pattern observed between (4d) and (3a).

(4) a. Zhāng Sān hái zài túshūguǎn.[continuative]
Zhang San HAI in library
'Zhang San is still in the library.'

b. Mìxiēgēn-hú hái zài Měiguó.[marginal]
Michigan-lake HAI in the U.S.
'Lake Michigan is still in the U.S.'

c. Shànghǎi bǐ Běijīng hái lěng.[scalar]
Shanghai BI Beijing HAI cold
'Shanghai is even colder than Beijing.'

d. Zhāng Sān xǐhuān kāfēi, hái xǐhuān chá.[additive]
Zhang San like coffee HAI like tea
'Zhang San likes coffee, and also likes tea.'

In line with this assumption, we adopt [2]'s perspective, positing that the additive *hái/noch* scopes over a speech act operator, *ANSWER* (cf. [30]), and provide its semantics in (5). According to (5), *hái* combines with an anaphoric proposition $p*$, an argument proposition p, and the speech act operator R.

(5) $[\![\text{hái}]\!] = \lambda p^*.\lambda p.\lambda R : p^* < p\&R(p^*).R(p)$

Hái asserts that its prejacent is presented as an answer to the question under discussion (QUD, [45])—$R(p)$. It presupposes that $p*$ precedes p in the order of mention (cf. [28]), which represents the precedence relation in discourse—$p* < p$, and that $p*$ is offered as an answer to the QUD—$R(p*)$.

Applying (5) to (4d), we propose that (4d) responds to an (explicit or implicit) QUD, as illustrated in (6a). The Logical Form (LF) of (4d) is represented in (6b). The additive *hái* in (4d) does not alter the at-issue meaning of its prejacent; instead, it contributes an additivity presupposition (PSP), as shown in (7).

(6) a. QUD: What does Zhang San like drinking?
b. LF: [$_{CP}$ *ANSWER* [$_{TP}$ Zhang San$_i$ likes coffee]] [$_{CP}$ *hái* [$_{CP}$ *ANSWER* [$_{TP}$ *pro*$_i$ likes tea]]]

(7) [[Zhang San hái likes tea]] =

a. At-issue: Zhang San likes tea.

b. PSP: An earlier-mentioned proposition in the discourse ("Zhang San likes coffee") contributes an answer to the QUD in (6a).

2.2 Focus Marker *Shì*

As indicated by the minimal pair in (8), the focus marker *shì* conveys an exhaustivity inference, resulting in the semantic contradiction seen in (8b).[1] Furthermore, (9) clarifies that this inference is a presupposition.

(8) a. Zhāng Sān lái-le, Lǐ Sì yě lái-le.
Zhang San come-ASP Li Si also come-ASP
'Zhang San came, and Li Si also came.'

b. #Shì Zhāng Sān lái-le, Lǐ Sì yě lái-le.
SHI Zhang San come-ASP Li Si also come-ASP
'It is Zhang San who came, and Li Si also came.'

(9) Negating *zhǐ* 'only' and *shì*:

a. Zhāng Sān hē-le kāfēi, dàn bù zhǐ hē-le kāfēi.
Zhang San drink-ASP coffee but not only drink-ASP coffee
'Zhang San drank the coffee, but he not only drank the coffee.'

b. #Zhāng Sān hē-le kāfēi, dàn bù shì hē-le kāfēi.
Zhang San drink-ASP coffee but not SHI drink-ASP coffee
'Zhang San drank the coffee, but it was not coffee that he drank.'

Just as [52] argues for English *only* and *it*-clefts, the at-issue and not-at-issue contents of Mandarin *zhǐ* 'only' and *shì* are reversed (see also [14]). *Shì*(p) asserts the prejacent p, whereas *zhǐ*(p) presupposes it. Conversely, *shì*(p) presupposes the exhaustivity inference, while *zhǐ*(p) asserts it.

Based on these empirical observations, we adopt the semantics of *shì* as proposed by [55], which builds on the work of [14, 52]. In (10), *shì* combines with a proposition p, asserting p while presupposing that no true answer is strictly stronger than p. This implies that p represents the maximal true answer to the QUD.

(10) $[\![shì]\!] = \lambda p : \forall q \in QUD[(q >_S p) \rightarrow \neg q].p$

[1] *Shì* is also homophonous and homographous with the copula. We propose that these two uses are synchronically distinct lexical items. For further discussion on the focus marker *shì*, see [14, 41, 55].

Using (11a) as an example, we propose that it addresses the QUD in (11b). The *shì* in (11a) asserts its prejacent while presupposing that it constitutes the maximal true answer to the QUD, as shown in (12).

(11) a. Zhāng Sān shì hē-le kāfēi.
Zhang San SHI drink-ASP coffee
'It was coffee that Zhang San drank.'

b. QUD: What did Zhang San drink?

(12) [[11a]] =

a. At-issue: Zhang San drank the coffee.

b. PSP: "Zhang San drank the coffee" contributes the maximal true answer to the QUD in (11b).

2.3 Háishì

With the semantics of *hái* and *shì* established, we propose that *háishì* involves the nesting of these particles, with *shì* falling under the scope of *hái*. Accordingly, the semantics of *háishì* is provided in (13). Like *shì*, *háishì* presupposes that its prejacent is the maximal true answer to the QUD. Since *hái* takes a wider scope than *shì*, its additivity presupposition signals that another maximal true answer has been previously mentioned in the discourse.

(13) $[\![\text{hái}[\text{shì } \psi]\,]\!]$

a. At-issue: $[\![\psi]\!]$.

b. PSPs: (i) $[\![\psi]\!]$contributes the maximal true answer to the QUD;

(ii) Another earlier-mentioned proposition contributes the maximal true answer to the QUD.

As shown in (13), the semantics of *háishì* proposed here is not related to disjunction. We take a further step by suggesting that φ *háishì* ψ represents the juxtaposition of φ and *háishì* ψ.[2] Following the literature, we propose a null *J*(unction) head as a set-forming operator ([1, 11, 49- 50, 53- 54], among others), as illustrated in (14). Based on the above assumptions, φ *háishì* ψ is interpreted as the *J* head collecting φ and *háishì* ψ into a set, as shown in (15).

(14) $[\![Jx_1, \ldots, x_n]\!] = \{[\![x_1]\!], \ldots, [\![x_n]\!]\}$

(15) $[\![\varphi \textit{ háishì } \psi\,]\!] = [\![J\, \varphi, \textit{háishì } \psi\,]\!]$
a. At-issue: $\{[\![\varphi]\!], [\![\psi]\!]\}$
b. PSPs: (i) $[\![\varphi]\!]$ contributes the maximal true answer to the QUD;
(ii) $[\![\psi]\!]$ contributes the maximal true answer to the QUD.

(15) articulates that the at-issue content of φ *háishì* ψ forms a set. The contribution of *háishì* signals that the juxtaposed propositions are mutually exclusive, implying that they cannot both be the maximal true answer simultaneously.

2 For convenience, we will continue to refer to it as disjunctive *háishì*, as seen in (3a), throughout the paper.

2.4 *Háishì* in Interrogatives

With the semantics of φ *háishì* ψ established in (15), the next step is to compositionally derive alternative questions involving *háishì*. The question operator we adopt, following [29], is straightforward and presented in (16).[3] This operator simply lets alternatives pass through. By combining (15) and (16), we can interpret interrogatives with *háishì*, as demonstrated in (17).

(16) Q operator: $[\![Q[\alpha]]\!] = [\![\alpha]\!]$

(17) $[\![Q[\varphi$ *háishì* $\psi]]\!] = [\![\varphi,$ *háishì* $\psi\ ]\!]$
a. At-issue: $\{\varphi, \psi\}$
b. PSPs: (i) $[\![\varphi]\!]$ contributes the maximal true answer to the QUD;
(ii) $[\![\psi]\!]$ contributes the maximal true answer to the QUD.

(17) follows the standard analysis of alternative questions in the literature (see [6], and references therein). According to this framework, an alternative question serves as a strategy for inquiring about an overarching *wh*-QUD, effectively inducing domain restriction for that QUD. Taking (1a), repeated as (18a), as an example, we assume it addresses the QUD presented in (18b). (18a) provides two possible answers—*coffee* and *tea*—to this QUD, presupposing that these two answers are mutually exclusive.[4]

(18) a. Zhāng Sān xǐhuān kāfēi háishì chá?
Zhang San like coffee HAISHI tea
'Does Zhang San like coffee or tea?'

b. QUD: What does Zhang San like drinking?

Furthermore, another intriguing fact can be elucidated, which, to my knowledge, has not received adequate attention and explanation. As shown in (19), another focus marker *shì* can optionally occur in alternative questions without altering the meaning. Our analysis readily extends to this phenomenon. *Háishì* presupposes that another proposition with exhaustive focus has been mentioned in the discourse. The presence of *shì* serves as an overt indication of the double cleft structure, while its absence is straightforward, as its contribution is consistently fulfilled by the additivity of *hái*.

(19) Zhāng Sān (***shì***) xǐhuān kāfēi háishì chá?

---

[3] It is also possible to employ a question operator that encodes the exhaustivity presupposition, as suggested in [44], to capture the complete semantic properties of alternative questions. However, since the primary focus of this paper is on the contribution of *háishì*, we will leave this option for future exploration.

[4] An anonymous reviewer suggested that the QUD addressed by (18a) could also be phrased as "Which drink does Zhang San prefer?" We agree with this intuition. According to [37], the questions in (i) share the same underlying representations. Singular *which*-questions carry a uniqueness presupposition (cf. [10]). Thus, the interpretation proposed by the reviewer effectively lifts the mutual exclusivity presupposition to the QUD. *Háishì* is presuppositionally congruent to this QUD and induces explicit domain restriction.(i) a. Does Zhang San like coffee or tea?b. Which drink does Zhang San like?*the domain of drinks restricted to coffee and tea*

Zhang San SHI like coffee HAISHI tea
'Does Zhang San like coffee or tea?'

Our analysis finds cross-linguistic support. For instance, [20] independently suggests that in Yoruba (Niger-Congo, Nigeria), the single focus particle *ni* following the disjunction functions as a CLEFT operator ([8]). This focus particle combines with each disjunct through pointwise function application, giving rise to the mutual exclusivity presupposition. Additionally, it's noteworthy that the pitch accents in each disjunct of English alternative questions have been proposed to indicate an overarching QUD as well ([19, 34]).

3 Alternative Questions

The proposed semantics and syntax of *háishì* contribute significantly to the discussion of alternative questions in the literature. Specifically, our analysis (i) supports the idea that alternative questions carry a mutual exclusivity presupposition, (ii) explains the distributional contrast between *háishì* and *huòzhě*, and (iii) unifies the syntactic strategies for constructing Mandarin alternative questions.

3.1 Mutual Exclusivity Presupposition

The question of whether alternative questions carry a mutual exclusivity presupposition is debated in the literature. Initially, it appears that they do not, as illustrated in (20), where each disjunct (a-b) and both together (c) can be acceptable responses to the question.

(20) Did Alfonso or Joanna bring a salad?

a. Alfonso did.

b. Joanna did.

c. They both did.

There are two opposing views on this issue. Some scholars, like [16, 27], argue that responses like (20c) qualify as answers, suggesting that alternative questions do not carry a mutual exclusivity presupposition. In contrast, others, including [1, 13, 11], contend that such responses do not answer the question but instead engage with the mutual exclusivity presupposition that the question encodes.

[5] shows that a presuppositional account provides better predictions for embedded contexts. For example, in the scenario presented in (21), the embedded alternative question in (21a) is infelicitous. This infelicity suggests that mutual exclusivity is indeed at play in alternative questions.

(21) Scenario: Suppose Henry knows that the plan was originally for either Alfonso or Joanna to make a salad (they were supposed to decide between them which would do it). Henry learns that Alfonso and Joanna were considering collaborating on the salad, so maybe both of them could bring it.

a. #Henry wondered whether Alfonso or Joanna would bring the salad.

b. Henry wondered whether Alfonso, Joanna, or both Alfonso and Joanna would bring the salad.
[5, pp. 378–379]

In this regard, Mandarin *háishì* offers morpho-syntactic evidence that bolsters the presuppositional camp, contributing to the resolution of the cross-linguistic debate on mutual exclusivity in alternative questions.

3.2 Distributional Contrast

Now we delve into the distributional contrast between *háishì* and *huòzhě*, addressing two key sub-questions: (i) Why is *huòzhě* not used in alternative questions? (ii) Why is *háishì* not used in polar questions or declaratives?

To answer the question (i), we first motivate the semantics of *huòzhě*. Following the literature, disjunction can be understood as existential quantification over an explicitly given domain containing two or more elements ([48], among many others). While (22a) serves as a sentence-level disjunction, we can extend this analysis following [38] by generalizing *huòzhě* to conjoin different types of constituents, as demonstrated in (22b).[5]

(22) a. $[\![huòzhě]\!] = \lambda T_{\langle t,t \rangle}.\exists p \in T\big[p\big]$
b. $[\![huòzhě]\!] = \lambda P_{\langle e,t \rangle}.\lambda Q_{\langle e,t \rangle}.\exists x \in P\big[Q(x)\big]$

This is further supported by two additional uses of *huòzhě*. As shown in (23), *huòzhě* also functions as an indefinite pronoun and a possibility modal, both of which have been analyzed in the literature as instances of existential quantification.

(23) a. Měi-tiān qīngchén dōu yǒu xǔduō rén zài gōngyuán duànliàn,
every-day morning DOU have many people in park exercise
huòzhě pǎobù, huòzhě dǎ-quán, huòzhě zuò-cāo.
HUOZHE jog HUOZHE practice-martial-arts HUOZHE do-calisthenics
'Every morning, many people are exercising in the park; some are jogging, some are practicing martial arts, and some are doing calisthenics.'([32, p. 284])
b. Nǐ gǎnkuài zǒu, huòzhě hái néng gǎnshàng mòbān-chē.
You quickly walk might still can catch last-bus
'Hurry up and leave; maybe you can still catch the last bus.'([32, p. 283])

As we argue, *huòzhě* disjunction is invariably existentially closed. This closure results in a sentence with *huòzhě* collapsing into a single proposition before being processed by the Q operator. Such semantics are fundamentally incompatible with alternative questions, which require more than one alternative proposition. Consequently, *huòzhě* cannot be utilized in alternative questions.

Now let's address the question (ii). We contend that a necessary condition for a sentence to qualify as a polar question or declarative is that the cardinality of its denotation equals one (cf. [5]). As shown in (15), φ *háishì* ψ constitutes a non-singleton set whose

[5] The precise semantics of *huòzhě* is not central to our analysis, provided that *huòzhě* disjunction is existentially closed before selection by the Q operator. For similar analyses, see [12, 15].

cardinality is evidently greater than one. Thus, we must introduce certain operators to transform this set into a single proposition.

(24) provides evidence that Mandarin exhibits zero conjunction, which could motivate the presence of a covert universal quantifier ([29]). However, given that *háishì* carries a mutual exclusivity presupposition, the introduction of such an operator would lead to a semantic conflict, as illustrated in (25).

(24) Zhāng Sān chàng-le gē, (bìngqiě) tiào-le wǔ.
Zhang San sing-ASP song and dance-ASP dance
'Zhang San sang songs and danced.'

(25) * $[\![\forall[\varphi$ *háishì* $\psi]]\!]$
At-issue: $\{\forall p[p \in \{[\![\varphi]\!], [\![\psi]\!]\} \rightarrow p = 1]\}$
PSPs: (i) $[\![\varphi]\!]$ contributes the maximal true answer to the QUD;
(ii) $[\![\psi]\!]$ contributes the maximal true answer to the QUD.

Furthermore, zero disjunction is absent in natural languages ([42, 49, 53, 54] among others). This implies that a covert existential quantifier ([29]) cannot be invoked without morpho-syntactic support. This independent cross-linguistic evidence clarifies why *háishì* is not used in polar questions or declaratives, as no operators can fulfill this role. In contrast, *huòzhě* functions as an expression of existential quantification, thereby permitting its use in these contexts.

3.3 Alternative Questions with and Without *Háishì*

In our framework, the two strategies for constructing Mandarin alternative questions converge into a single approach: φ *háishì* ψ is identified as a marked juxtaposition (MJ). However, using bare juxtaposition (BJ) to derive alternative questions is not as flexible as that of MJ.

First, monosyllables normally cannot form alternative questions using BJ, as illustrated by the contrast in (26). Conversely, MJ imposes no such syllabic restrictions, as demonstrated in (27).[6]

(26) a. *Zhāng Sān chī fàn, miàn?
Zhang San eat rice noodle

b. Zhāng Sān chī mǐfàn, miàntiáo?
Zhang San eat rice noodle
'Does Zhang San eat rice or noodles?'

(27) a. Zhāng Sān chī fàn háishì miàn?
Zhang San eat rice HAISHI noodle

b. Zhāng Sān chī mǐfàn háishì miàntiáo?
Zhang San eat rice HAISHI noodle
'Does Zhang San eat rice or noodles?'

[6] We challenge the generalization put forth by [22–23] that juxtaposed elements in alternative questions must exhibit partial phonological identity, as evidenced by example (26b). This observation aligns with note 8 in [25].

Second, syntactic categories and positions play a crucial role for BJ but are not restrictive for MJ. For instance, BJ of adverbial phrases typically fails to form alternative questions, whereas MJ succeeds, as shown in (28). The minimal pair in (29) further illustrates that BJ in the subject position results in ill-formedness, highlighting a subject-object asymmetry.[7]

(28) a. *Zhāng Sān kāixīn-de shāngxīn-de lái-le?
Zhang San happily sadly come-ASP
b. Zhāng Sān kāixīn-de háishì shāngxīn-de lái-le?
Zhang San happily HAISHI sadly come-ASP
'Did Zhang San come happily or sadly?'

(29) a. *Zhāng Sān, Lǐ Sì lái-le?
Zhang San Li Si come-ASP
b. Zhāng Sān háishì Lǐ Sì lái-le?
Zhang San HAISHI Li Si come-ASP
'Did Zhang San or Li Si come?'

Third, a contrast exists between BJ and MJ concerning the licensing of *wh*-indefinites, as highlighted by [25, 26]. Mandarin *wh*-phrases, such as *shéi* 'who', are recognized for their non-interrogative uses (cf. [9, 31], among many others). As illustrated in (30), *shéi* in BJ can only be interpreted interrogatively, whereas in MJ, *shéi* is interpreted exclusively as existential.

(30) a. Nǐ dǎ-le shéi mà-le shéi?
you hit-ASP who scold-ASP who
✓Interrogative: 'Who did you hit and who did you scold?'
* Existential: 'Did you hit someone or scold someone?'
b. Nǐ dǎ-le shéi háishì mà-le shéi?
you hit-ASP who HAISHI scold-ASP who
* Interrogative: 'Who did you hit and who did you scold?'
✓Existential: 'Did you hit someone or scold someone?'

Fourth, we propose a semantic hierarchy for BJ to be interpreted as alternative questions. Drawing from lexical semantics, encyclopedic knowledge, and other factors, we can identify at least three types of relations in BJ: opposite > contrastive > compatible. Here, '>' signifies the potential for interpretation as alternative questions, with the leftmost relation being the most mutually exclusive.

The opposite relation pertains to the juxtaposition of components that represent two polarities of a single property. For instance, as illustrated in (31), pairs like *male* and *female*, or *on* and *off*, are widely recognized as two opposing categories, satisfying the mutual exclusivity requirement of alternative questions. Consequently, this type of BJ is usually interpreted seamlessly as an alternative question.

[7] The ill-formedness of (29a) pertains specifically to the alternative question reading. (29a) can still be felicitously interpreted as a polar question, wherein the juxtaposed elements imply a zero conjunction.

(31) Opposite:

a. Zhāng Sān nán-de nǚ-de?
Zhang San man-DE woman-DE
'Is Zhang San male or female?'
b. Dēng kāi-zhe guān-zhe?
light open-ASP close-ASP
'Is the light/are the lights on or off?

The contrastive relation does not exhibit the sharp conflict seen in the opposite relation. In (32), the BJ is not lexically incompatible; one can easily envision a context in which Zhang San eats both rice and noodles at the same meal, or a TV excels in both color and sound quality. The mutually exclusive inference in (32a) arises from encyclopedic knowledge, as it is common for Chinese people to consume only one type of staple food per meal. The phonological and structural patterns in (32b) highlight its contrast.

(32) Contrastive:
a. Zhāng Sān chī mǐfàn, miàntiáo?
Zhang San eat rice noodle
'Does Zhang San eat rice or noodles?'
b. Zhè tái diànshì sècǎi hǎo yīnzhì hǎo?
this CL TV color good sound-quality good
'Does this TV have good color or good sound quality?'

The minimally modified data in (33) from (32) becomes infelicitous in an out-of-the-blue context. This compatible relation indicates that, in such cases, BJ is generally understood as non-mutually exclusive.

(33) Compatible: a. ??Zhāng Sān chī fàn, hē tāng?
Zhang San eat rice drink soup
Intended: 'Does Zhang San eat rice or have soup?'
b. ??Zhè tái diànshì sècǎi hǎo yīnzhì huài?
this CL TV color good sound-quality bad
Intended: 'Does this TV have good color or poor sound quality?'

We propose that the degradation is pragmatically triggered by severe QUD accommodation. The infelicity of (33) arises from the difficulty in accommodating a *wh*-QUD suggesting that *eating rice* and *having soup*, or *good color* and *poor sound quality*, are mutually exclusive answers. However, if the context is sufficiently supportive, facilitating the accommodation of such a QUD, the examples in (33) become felicitous, as evidenced in (34).

(34) a. Context: Zhang San was being quite naughty and refused to eat properly at dinner. His mother insisted that even if he wasn't hungry, he needed to eat something. She told him he should either have some rice or some soup; otherwise, he would be hungry later. His mother said:
Nǐ chī fàn, hē tāng?
you eat rice drink soup
'Do you eat rice or have soup?'

b. Context: Zhang San went to the mall to buy a TV. The salesperson informed him that there are currently two types of TVs available: one with excellent color quality and the other with poor sound quality. Zhang San pointed to a TV in front of him and asked the salesperson:
Zhè tái diànshì sècǎi hǎo yīnzhì huài?
this CL TV color good sound-quality bad
'Does this TV have good color or poor sound quality?'

Needless to say, these data require a comprehensive solution, which unfortunately exceeds the scope of this paper. However, what is crucial for our analysis here is that the mutual exclusivity requirement plays a key role in Mandarin alternative questions. On one hand, this property is what sets *háishì* apart from *huòzhě*, making the former to be restrictively distributed in alternative questions. On the other hand, the semantic hierarchy of BJ demonstrates that the more mutually exclusive the components are, the more readily they can be licensed as alternative questions. Additionally, the lexical semantics of *háishì* accounts for why sentences in (33), when framed with *háishì*, remain more felicitous even in an out-of-the-blue context.

4 A Uniform Analysis of Three *Háishì*'s

After elucidating the semantics of disjunctive *háishì*, this section explores how the decompositional analysis unifies the meanings of its three distinct uses.

4.1 Unified Semantics for *Hái*

In Sect. 2.1, we demonstrated that Mandarin *hái* has numerous uses. Following [18], which builds on [3], we propose a unified semantics for *hái*, as outlined in (35).

(35) $[\![hái]\!] = \lambda S.\lambda x^*.\lambda x.\lambda P_{<x,t>} : x^* <_S x \& P(x^*).P(x)$

(35) states that *hái* always combines with a scale *S*, an anaphoric element *x**, an argument *x*, and a predicate *P*. It does not alter the assertion that the predicate is true of *x*—*P*(*x*). Instead, it presupposes that *x** precedes *x* on the scale, and that the predicate is true of x^*—$x^* <_S x \,\&\, P(x^*)$.

Different scales give rise to various uses of *hái*. As previously stated, additive *hái* relies on the order of mention, as reiterated in (36).

(36) $[\![hái_{\text{additive}}]\!] = \lambda p^*.\lambda p.\lambda R : p^* < p \& R(p^*).R(p)$

In the case of continuative *hái*, the relevant scale is temporal order '$<$', which denotes the precedence relation on time intervals. The relation '$\prec$', representing immediate precedence, is a subset of '$<$' and is necessary to capture the continuative inference. The semantics of continuative *hái* is outlined in (37). Consider example (38).[8]

(37) $[\![hái]\!] = \lambda t^*.\lambda t.\lambda P_{<i,t>} : t^* \prec t \& P(t^*).P(t)$

[8] Since the specific semantics of other uses of *hái* are not central to this paper, we will not elaborate on the analysis here. Interested readers can refer to [3] for further details.

(38) [[Zhang San is ***hái*** in the library]]=
a. At-issue: Zhang San is in the library.
b. PSP: Zhang San was in the library earlier.

4.2 Continuative *Háishì*

We propose that the meaning of continuative *háishì* arises from the interaction between continuative *hái*, as shown in (37), and the focus marker *shì*, illustrated in (10). The semantics of continuative *háishì* is given in (39). Notably, the temporal order is anchored not in events but in speech acts. The presupposition of *háishì* indicates that multiple sister QUDs have been proposed within the discourse tree (cf. [7]). The prejacent of continuative *háishì* contributes the maximal true answer to the current QUD and its preceding sister QUD.

(39) $[\![hái[shì\,\psi]_{\text{continuative}}]\!]$ =
a. At-issue: $[\![\psi]\!]$.
b. PSPs: (i) $[\![\psi]\!]$ contributes the maximal true answer to the current QUD;
(ii) $[\![\psi]\!]$ contributes the maximal true answer to the preceding sister QUD in the discourse tree.

Consider (40a). We assume it addresses the sister QUDs presented in (40b-c), where QUD_2 precedes QUD_1 in the discourse tree.[9] The meaning components are detailed in (41).

(40) a. Zhāng Sān háishì zài túshūguǎn.
Zhang San HAISHI in library
'Zhang San is still in the library.'
b. QUD_1: Where is Zhang San now?
c. QUD_2: Where was Zhang San before?

(41) [[40a]] =
a. At-issue: Zhang San is in the library.
b. PSPs: (i) "Zhang San is in the library" contributes the maximal true answer to the QUD_1;
(ii) "Zhang San was in the library" contributes the maximal true answer to the QUD_2.

At first glance, the semantics of continuative *háishì* and continuative *hái* appear similar. Mandarin reference grammar notes that they can often be interchanged (e.g., [32, p. 254]), as shown in (42). However, this apparent interchangeability arises from the coincidence that the sequence of temporal contrastive topics in the QUDs aligns with the event time. Our analysis posits that the continuative meaning of *hái* is anchored in event time, which accounts for the infelicity of (43a). In contrast, the temporal inference of *háishì* pertains to QUDs, thereby predicting that (43b) is well-formed.

[9] Since Mandarin does not morphologically encode tense information in the verb, *Zhāng Sān zài túshūguǎn* 'Zhang San be in the library' is felicitous in answering both QUD_1 and QUD_2.

(42) Zhāng Sān hái / háishì zài túshūguǎn.
Zhang San HAI HAISHI in library
'Zhang San is still in the library.'

(43) a. #Zhāng Sān xiàwǔ zài túshūguǎn, zǎoshang hái zài túshūguǎn.
Zhang San afternoon in library morning HAI in library
'Zhang San was in the library in the afternoon and was still there in the morning.'
b. Zhāng Sān xiàwǔ zài túshūguǎn, zǎoshang háishì zài túshūguǎn.
Zhang San afternoon in library morning HAISHI in library
'Zhang San was in the library in the afternoon and was also there in the morning.'

4.3 Suggestive *Háishì*

We propose that the meaning of suggestive *háishì* is equivalent to disjunctive *háishì*, as illustrated in (44).

(44) $[\![hái[shì\psi]_{\text{suggestive}}]\!] = [\![hái[shì\psi]_{\text{disjunctive}}]\!] =$
a. At-issue: $[\![\psi]\!]$.
b. PSPs: (i) $[\![\psi]\!]$ contributes the maximal true answer to the QUD;
(ii) An earlier-mentioned proposition contributes the maximal true answer to the QUD.

According to this framework, (45B) is interpreted as (46). This sentence directly addresses the QUD posed in (45A): "Where shall we go?" (45B) asserts its prejacent while presupposing that it represents the maximal true answer and that another maximal true answer has already been proposed. The mutual exclusivity presupposition effectively negates the proposal made in (45A), conveying a sense of refusal.

(45) A: Where shall we go? How about Shanghai?
B: Wǒmen háishì qù Běijīng ba!
we HAISHI go Beijing SFP
'Let's go to Beijing instead!'

(46) $[\![45\text{B}]\!] =$
a. At-issue: We go to Beijing.
b. PSPs: (i) "We go to Beijing" contributes the maximal true answer to the QUD;
(ii) An earlier-mentioned proposition ("going to Shanghai") contributes the maximal true answer to the QUD.

The semantics we propose successfully predicts that (45B) cannot be used to make an initial suggestion, as demonstrated in (47).

(47) A: Where shall we go?
B: #Wǒmen háishì qù Běijīng ba!
we HAISHI go Beijing SFP
'Let's go to Beijing instead!'

It is quite intriguing that *háishì* can exhibit such flexible meanings depending on its interaction with different speech acts. Due to the mutual exclusivity presupposition of *háishì*, the propositions it associates can only coexist in questions, where the speaker prompts the listener to choose one among these alternatives. In assertions, however, the speaker cannot assert multiple mutually exclusive propositions. For the use of *háishì* in assertions to be valid, the propositions it links must be expressed by different speakers, with the prejacent of *háishì* negating all other alternatives.

5 Conclusion and Remaining Issues

In this paper, we propose a decompositional analysis of Mandarin *háishì*. We argue that *háishì* is not an interrogative disjunctor but rather an adverb taking scope over a full CP; thus, φ *háishì* ψ represents the juxtaposition of φ and *háishì* ψ. The nesting of additive *hái* and the focus marker *shì* contributes to the mutual exclusivity presupposition of *háishì*. This proposal seems to be on the right track, as it not only addresses the distributional contrast between *háishì* and *huòzhě* but also unifies the syntactic strategies for constructing Mandarin alternative questions and provides a uniform analysis of the three uses of *háishì*.

Going forward, several questions and open issues arise. First, our analysis predicts that *háishì* juxtaposition in alternative questions is fundamentally based on CP clauses.[10] The apparent size of the subclause may result from applying an ellipsis mechanism, such as the move-and-delete approach proposed by [35, 36]. However, [12, 15] strongly oppose the clausal disjunct analysis, presenting substantial evidence against it (see Appendix A of [15]).

While a comprehensive rebuttal of [15]'s arguments is beyond the scope of this paper, it is noteworthy that the sentence-final particle *ne*, which is within the CP domain ([13, 39, 40], among many others), can appear in individual juxtaposed elements, as illustrated in (48). Moreover, the focus marker *shì* is disallowed in clauses lacking the CP domain ([14]), as exemplified in (49). The optional presence of the focus marker *shì* to the left of a *háishì* juxtaposition further supports our double CP analysis.

(48) Zhāng Sān xǐhuān kāfēi ***ne*** háishì chá ***ne***?
Zhang San like coffee SFP HAISHI tea SFP
'Does Zhang San like coffee or tea?'

(49) *Shì* disallowed in subject control complement ([14, p. 19]):
Wǒ xiǎng [{*shì, zhǐ} hē [kāfēi]$_F$].
I want SHI only drink coffee
* ≈ 'I want [for it to be coffee that I drink].'
✓ 'I want [to only drink [coffee]$_F$].'

[10] [24–26] propose that Mandarin alternative questions consistently involve the disjunction of full clauses. [24, p. 250] refer to these structures as "full-size, bi-clausal sources," while [25] describes them as being of "TP/IP" size.

Additionally, the detailed distinction between BJ and MJ in Mandarin alternative questions necessitates a more thorough description and explanation. The semantics of Mandarin alternative questions and the three uses of *háishì* still require comprehensive articulation and formalization.

Acknowledgments. I am grateful to my advisor, Mingming Liu, for his support and suggestions. I also appreciate the helpful comments from the reviewers of this paper.

Disclosure of Interests. The author has no competing interests to declare that are relevant to the content of this article.

References

1. Alonso-Ovalle, L.: Disjunction in alternative semantics. PhD Dissertation, University of Massachusetts Amherst (2006)
2. Beck, S.: Discourse related readings of scalar particles. In: Moroney, M., Little, C.R., Collard, J., Burgdorf, D. (eds.) Proceedings of Semantics and Linguistic Theory 26, pp. 142–165 (2016). https://doi.org/10.3765/salt.v26i0.3783
3. Beck, S.: Readings of scalar particles: noch/still. Linguist. Philos. **43**(1), 1–67 (2020). https://doi.org/10.1007/s10988-018-09256-1
4. Belnap, N., Steel, T.: The logic of questions and answers. Yale University Press, New Haven (1976). https://doi.org/10.2307/2272838
5. Biezma, M., Rawlins, K.: Responding to alternative and polar questions. Linguist. Philos. **35**, 361–406 (2012). https://doi.org/10.1007/s10988-012-9123-z
6. Biezma, M., Rawlins, K.: Alternative questions. Lang. Linguist. Compass **9**(11), 450–468 (2015). https://doi.org/10.1111/lnc3.12161
7. Büring, D.: On D-trees, beans, and B-accents. Linguist. Philos. **26**, 511–545 (2003). https://doi.org/10.1023/A:1025887707652
8. Büring, D., Kriz, M.: It's that, and that's it! Exhaustivity and homogeneity presuppositions in clefts (and definites). Semant. Pragmatics **6**, 1–29 (2013). https://doi.org/10.3765/sp.6.6
9. Chao, Y.: A grammar of spoken Chinese. University of California Press (1968)
10. Dayal, V.: Locality in WH Quantification. Kluwer Academic Publishers, Dordrecht (1996)
11. den Dikken, M.: Either-float and the syntax of coordination. Nat. Lang. Linguist. Theory **24**, 689–749 (2006). https://doi.org/10.1007/s11049-005-2503-0
12. Erlewine, M.: Alternative questions through focus alternatives in Mandarin Chinese. In: Beltrama, A., Chatzikonstantinou, T., Lee, J., Pham, M., Rak, D. (eds.) Proceedings of the 48th Meeting of the Chicago Linguistic Society, pp. 221–234 (2014)
13. Erlewine, M.: Low sentence-final particles in Mandarin Chinese and the final-over-final constraint. J. East Asian Linguis. **26**, 37–75 (2017). https://doi.org/10.1007/s10831-016-9150-9
14. Erlewine, M.: Mandarin Shì clefts and the syntax of discourse congruence. Manuscript. https://ling.auf.net/lingbuzz/005176 (2020)
15. Erlewine, M.: Interrogative and standard disjunction in Mandarin Chinese. Manuscript. https://ling.auf.net/lingbuzz/008015 (2024)
16. Groenendijk, J., Stokhof, M.: Studies on the semantics of questions and the pragmatics of answers. PhD dissertation, University of Amsterdam (1984)

17. Haspelmath, M.: Coordination. In: Shopen, T. (ed.) Language typology and syntactic description, vo. 2, pp. 1–51. Cambridge University Press (2007)
18. He, R.: Scalar or additive: A unified semantics for Mandarin Hái. Talk given at the 30th Annual Conference of International Association of Chinese Linguistics, Seoul, Korea (2024)
19. Hoeks, M.: The role of focus marking in disjunctive questions: A QUD-based approach. In: Rhyne, J., Lamp, K., Dreier, N., Kwon, C. (eds.) Proceedings of Semantics and Linguistic Theory 30, pp. 654–673 (2020). https://doi.org/10.3765/salt.v30i0.4852
20. Howell, A.: A Hamblin semantics for alternative questions in Yoruba. In: Bade, N., Berezovskaya, P., Schöller, A. (eds.) Proceedings of Sinn und Bedeutung 20, pp. 359–376 (2016)
21. Huang, C.-T.J.: Logical relations in Chinese and the theory of grammar. PhD dissertation, Massachusetts Institute of Technology (1982)
22. Huang, C.-T.J.: Hànyǔ zhèngfǎn wènjù de mózǔ yǔfǎ [A modular grammar of Chinese A-not-A questions]. Zhōngguó Yǔwén [Studies of the Chinese Language], 247–264 (1988)
23. Huang, C.-T.J.: Modularity and Chinese A-not-A questions. In: Georgopolous, C., Ishihara, R. (eds.) Interdisciplinary Approaches to Language, pp. 305–322. Dordret: Kluwer (1991). https://doi.org/10.1007/978-94-011-3818-5_16
24. Huang, C.-T.J., Li, Y.-H.A., Li, Y.: The syntax of Chinese. Cambridge University Press (2009).https://doi.org/10.1017/cbo9781139166935
25. Huang, R.-H.R.: Delimiting three types of disjunctive scope in Mandarin Chinese. In: Hu, C.-Y., Chiang, Y.-W.K. (eds.) University System of Taiwan working papers in linguistics, vol. 5, pp. 59–71 (2009)
26. Huang, R.-H.R.: Disjunction, coordination, and question: a comparative study. PhD Dissertation, National Taiwan Normal University (2010)
27. Karttunen, L.: Syntax and semantics of questions. Linguist. Philos. **1**, 3–44 (1977). https://doi.org/10.1007/BF00351935
28. Klein, W.: Time and again. In: Féry, C., Sternefeld, W. (eds.) Audiatur Vox Sapientiae: A Festschrift for Arnim von Stechow, pp. 267–286. Akademie, Berlin (2001)
29. Kratzer, A., Shimoyama, J.: Indeterminate pronouns: the view from Japanese. In: Otsu, Y. (ed.) Proceedings of the Tokyo conference on psycholinguistics 3, pp. 1–25. Tokyo: Hituzi Syobo (2002). https://doi.org/10.1007/978-3-319-10106-4_7
30. Krifka, M.: Embedding illocutionary acts. In: Roeper, T., Speas, M. (eds.) Recursion, Complexity in Cognition. Studies in Theoretical Psycholinguistics 43, pp. 125–155. Springer, Berlin (2014). https://doi.org/10.1007/978-3-319-05086-7_4
31. Li, Y.-H.A.: Indefinite Wh in mandarin Chinese. J. East Asian Linguis. **1**(2), 125–155 (1992). https://doi.org/10.1007/BF00130234
32. Lǚ, S.: Xiàndài hànyǔ bābǎi cí [800 words in Modern Chinese]. Beijing: Shāngwù Yìnshūguǎn [The Commercial Press] (1999)
33. Mauri, C.: The irreality of alternatives: towards a typology of disjunction. Stud. Lang. **32**, 28–55 (2008). https://doi.org/10.1075/sl.32.1.03mau
34. Meertens, E., Kutscheid, S., Romero, M.: Multiple accent in alternative questions. In: Espinal, M. T., Castroviejo, E., Leonetti, M., McNally, L., Real-Puigdollers, C. (eds.) Proceedings of Sinn und Bedeutung 23, vol. 2, pp. 179–196 (2019)
35. Merchant, J.: The syntax of silence: sluicing, islands, and the theory of ellipsis. Oxford: Oxford University Press (2001). https://doi.org/10.1093/oso/9780199243730.001.0001
36. Merchant, J.: Fragments and ellipsis. Linguist. Philos. **27**, 661–738 (2004). https://doi.org/10.1007/s10988-005-7378-3
37. Nicolae, A.: Alternative questions as strongly exhaustive WH-questions. In: Iyer, J., Kusmer, L. (eds.) Proceedings of the Forty-Fourth Annual Meeting of the North East Linguistic Society, vol. 2, pp. 65–78 (2014)

38. Partee, B., Rooth, M.: Generalized conjunction and type ambiguity. Meaning, use and interpretation of language, pp. 361–383. Berlin, Germany: de Gruyter (1983). https://doi.org/10.1515/9783110852820.361
39. Paul, W.: Why particles are not particular: sentence-final particles in Chinese as heads of a split CP. Studia Linguistica **68**, 77–115 (2014). https://doi.org/10.1111/stul.12020
40. Paul, W.: New perspectives on Chinese syntax. Berlin: De Gruyter Mouton (2015). https://doi.org/10.1515/9783110338775
41. Paul, W., Whitman, J.: Shi… de focus clefts in mandarin Chinese. Linguist. Rev. **25**(3–4), 413–451 (2008). https://doi.org/10.1515/TLIR.2008.012
42. Payne, J.: Complex phrases and complex sentences. In: Shopen, T. (ed.) Language Typology and Syntactic Description 2, Complex Constructions, Cambridge University Press (1985)
43. Rawlins, K.: (Un)conditionals: an investigation in the syntax and semantics of conditional structures. PhD Dissertation, University of California, Santa Cruz (2008)
44. Rawlins, K.: (Un)conditionals. Nat. Lang. Seman. **21**, 111–178 (2013). https://doi.org/10.1007/s11050-012-9087-0
45. Roberts, C.: Information structure: towards an integrated formal theory of pragmatics. Semant. Pragmatics **5**(6), 1–69 (1996/2012). https://doi.org/10.3765/sp.5.6
46. Rooth, M.: Association with focus. PhD Dissertation, University of Massachusetts Amherst (1985)
47. Rooth, M.: A theory of focus interpretation. Nat. Lang. Seman. **1**, 75–116 (1992). https://doi.org/10.1007/BF02342617
48. Rooth, M., Partee, B.: Conjunction, type ambiguity and wide scope "or". In: Flickinger, D. P., Macken, M., Wiegand, N. (eds.) Proceedings of the First West Coast Conference on Formal Linguistics, vol. 1. Linguistics Dept., Stanford University (1982)
49. Szabolcsi, A.: What do quantifier particles do? Linguist. Philos. **38**, 159–204 (2015). https://doi.org/10.1007/s10988-015-9166-z
50. Simons, M.: Dividing things up: the semantics of or and the modal/or interaction. Nat. Lang. Seman. **13**, 271–316 (2005). https://doi.org/10.1007/s11050-004-2900-7
51. Tsai, C.-Y.: Toward a theory of mandarin quantification. PhD Dissertation, Harvard University (2015)
52. Velleman, D., Beaver, D., Destruel, E., Bumford, D., Onea, E., Coppock, E.: It-clefts are IT (Inquiry Terminating) constructions. In: Chereches, A. (ed.) Proceedings of Semantics and Linguistic Theory 22, pp. 441–460. Ithaca, NY: CLC (2012). https://doi.org/10.3765/salt.v22i0.2640
53. Winter, Y.: Syncategorematic conjunction and structured meanings. In: Simons, M., Galloway, T. (eds.) Proceedings of Semantics and Linguistic Theory 5, pp. 387–404 (1995). https://doi.org/10.3765/salt.v5i0.2704
54. Winter, Y.: Flexible boolean semantics: Coordination, plurality and scope in natural language. PhD Dissertation, Utrecht University (1998)
55. Ye, S.: From maximality to bias: biased A-not-A questions in Mandarin Chinese. In: Rhyne, J., Lamp, K., Dreier, N., Kwon, C. (eds.) Proceedings of Semantics and Linguistic Theory 30, pp. 355–375 (2020). https://doi.org/10.3765/salt.v30i0.4826

Disjunctions of Universal Modals and Conditionals

Dean McHugh(✉)

Department of Philosophy and Institute for Logic, Language and Computation, University of Amsterdam, Amsterdam, The Netherlands
deanmchugh1@gmail.com

Abstract. This paper is concerned with disjunctions of universal modals, such as *You have to clean your room or you have to walk the dog*, and disjunctions of conditionals, such as *If Alice dances, Charlie will dance, or if Bob dances, Charlie will dance.* We aim to provide a uniform account of their surprising behaviour. Our proposal combines three independently attested components. Firstly, disjunction's dynamic effect, familiar from presupposition projection: when we evaluate a disjunction $A \vee B$, we typically interpret B assuming the negation of A, and optionally, also interpret A assuming the negation of B. Secondly, the fact that disjunctions often receive a conjunctive interpretation, familiar from free choice phenomena. Thirdly, that modal bases are restricted to local contexts, what Mandelkern has called 'bounded modality'.

1 Introduction

Remarkable things happen when disjunction joins universal modals and conditionals, as in (1) and (2).

(1) a. You must do clean your room or you must walk the dog.
 b. The keys must be in the drawer or John must have taken them.

(2) a. If Alice had come to the party, Charlie would have come. Or if Bob had comc, Charlic would havc comc.
 b. If Alice had come to the party, Charlie would have come. Or if Alice had come, Darius would have come.

We study these two cases of disjunction in tandem since we believe that a uniform explanation underlies their surprising behaviour.

Disjunctions of universal modals such as *Must A or must B*, first discussed by Geurts [18], behave in a way unexpected from a simple application of classical logic: they can be asserted even when both disjuncts are false. You must clean your room or you must walk the dog, but it is not true that you must clean your room (you may walk the dog and leave your room as it is), and it is not true that you must walk the dog (you may clean your room and keep the dog home). The keys must be in the drawer or John must have taken them, but it is not true that they must be in the drawer (for John might have taken them), and it

J. Yan et al. (Eds.): TLLM 2024, LNCS 15410, pp. 69–92, 2025.
https://doi.org/10.1007/978-3-031-86054-6_4

is not true that John must have taken them (for they might be in the drawer). Naturally, this violates classical logic. If both disjuncts are false, classical logic tells us that the disjunction as a whole must be false too.

Instead of the meaning we would expect from classical logic, *Must A or must B* somehow winds up meaning something akin to $Must(A \vee B)$. This is not how disjunctions of universal quantifiers usually behave.

(3) a. Everyone is in the kitchen or the garden.
b. $\not\equiv$ Everyone is in the kitchen or everyone is in the garden.

If some people are in the kitchen and the rest are in the garden, the first is true but the second false.[1] The problem becomes especially salient when we unpack the meaning of *must* according to a standard, Kratzerian semantics of modals.

(4) a. You must clean the kitchen or you must walk the dog.
b. $\not\equiv$ In every normatively best world you clean the kitchen, or in every normatively best world you walk the dog.

Our first goal is to account for this unexpected behaviour of disjunctions of universal modals.

Question 1. What are the truth-conditions of disjunctions of universal modals? In particular, why are they assertable even when both disjunctions are false?

Turning to disjunctions of conditionals, discussed by Woods [65], Geurts [18], and Khoo [30], we find an interesting contrast. When the antecedents are different—as in (2a), *If A, C or if B, C*—we readily perceive a conjunctive interpretation, with the sentence implying both disjuncts ($\rightsquigarrow$ denotes an intuitive felt inference, while remaining neutral on its precise status).

(2a) If Alice had come to the party, Charlie would have come. Or if Bob had come, Charlie would have come.
$\rightsquigarrow$ If Alice had come to the party, Charlie would have come.
$\rightsquigarrow$ If Bob had come, Charlie would have come.

However, when the antecedents are the same, as in (2b), the inference disappears.

[1] At least, the first *can* be true in this scenario; it also has a wide-scope reading on which it is equivalent to the second. This can be explained by the well-known scopal flexibility of disjunction, illustrated by (i) from Rooth and Partee [52].

(i) Mary is looking for a maid or a cook.

As Rooth and Partee observe, this has a reading suggested by the continuation "... but I don't know which", where the disjunction takes wide-scope, interpreted as "Mary is looking for a maid or Mary is looking for a cook".

Nonetheless, while (3a) has both narrow and wide scope readings, (3b) only has the wide scope reading.

(2b) If Alice had come to the party, Charlie would have come. Or if Alice had come, Darius would have come.
$\not\rightsquigarrow$ If Alice had come to the party, Charlie would have come.
$\not\rightsquigarrow$ If Bob had come, Charlie would have come.

The two consequents are logically consistent: none of the information provided rules out that Charlie and Darius may both come to the party. Thus we cannot point to logical incompatibility as a reason why the inference disappears here.

Nor is the contrast explained by an exclusive reading of disjunction. One might propose that an exclusive reading blocks us from inferring in (2b) that both conditionals are true, since this is incompatible with the exclusive inference, and that since we cannot choose which conditional to not infer, by symmetry we infer neither. The exact same, however, may be said for (2a). This account does not rule out an exclusive reading of (2a), whereas we in fact infer that both conditionals are true. The possibility of an exclusive reading of disjunction does not account for the fact that (2a) is read conjunctively and (2b) disjunctively.

Nor can we trace any difference between the conditionals to the fact that one of them contains some redundancy, unnecessarily repeating material. Both (2a) and (2b) are on a par in terms of how much redundancy they contain—the meaning of each can equivalently be expressed with greater brevity, say, like so.

(5) a. If Alice or Bob had come to the party, Charlie would have come.
b. If Alice had come to the party, Charlie or Darius would have come.

Interestingly, the conjunctive interpretation can even arise when the antecedents are distinct. For instance, (6) has a conjunctive interpretation.

(6) If you had taken the morning train, you would have arrived before lunch. Or if you had taken the afternoon train, you would have arrived after lunch.

The contrast is robust across the various forms conditionals may take. It occurs regardless of the conditional's tense.[2] The contrast also appears with conditional readings of conjunctions.[3]

[2] For example, in each case below, as in (2), the (a)-sentences typically imply both conditionals, while the (b)-sentences do not.

(i) a. If Alice goes, Charlie goes, or if Bob goes, Charlie goes.
b. If Alice goes, Charlie goes, or if Alice goes, Darius goes.

(ii) a. If Alice goes, Charlie will go, or if Bob goes, Charlie will go.
b. If Alice goes, Charlie will go, or if Alice goes, Darius go will go.

(iii) a. If Alice went, Charlie would go, or if Bob went, Charlie would go.
b. If Alice went, Charlie would go, or if Alice went, Darius would go.

[3] For recent analyses of conditional conjunctions see von Fintel and Iatridou [17], Starr [59], Keshet and Medeiros [29], and Kaufmann and Whitman [28], among others.

(7) a. Invite Alice and Charlie will go, or invite Bob and Charlie will go.
b. Invite Alice and Charlie will go, or invite Alice and Darius will go.

This shows that the preference for a conjunctive reading of *If A, C or if B, C* but a disjunctive reading of *If A, C or if A, D* is not specific to the conditional construction itself (say, the presence of *if*), but is rather a fact about conditional meaning broadly construed.

The conjunctive inference is cross-linguistically robust. Consider, for example, the following passage from the Book of Leviticus:

(8) And if a soul sin ... if he do not utter it, then he shall bear his iniquity.
Or if a soul touch any unclean thing ... he also shall be unclean, and guilty.
Or if he touch the uncleanness of man ... when he knoweth of it, then he shall be guilty. (Leviticus 5:1–3, King James Version, 1611).

This is most naturally read as a conjunction of conditionals. Cross-linguistically, a disjunction word links the clauses of Leviticus 5 in, for example, Mandarin Chinese *huò*, the original Hebrew *o*, Hungarian *vagy*, Icelandic *eða*, Māori *rānei*, Urdu *yâ*, Somali *ama*, Welsh *neu*, and Yoruba *tàbí*, suggesting that the conjunctive reading of disjunctions of conditionals is a cross-linguistically robust phenomenon.[4]

Our second goal is to capture this surprising behaviour of disjunctions of conditionals.

Question 2. What are the truth conditions of disjunctions of conditionals? In particular, why does *If A, C or if B, C* by default receive a conjunctive interpretation, while *If A, C or if A, D* does not?

There is an extensive literature on conditionals with disjunctive antecedents; in particular, on simplification of disjunctive antencedents—the inference from *if A or B, C* to *if A, C and if B, C*.[5] More recently, Khoo [31] and Klinedinst [32] consider the case of *if or if*-conditionals such as (9).[6]

---

[4] For sources see the Appendix.

[5] Among authors who argue for simplification's validity are Nute [47], Ellis, Jackson, and Pargetter [11], Warmbrōd [62], Fine [13], Starr [58], and Willer [64]. Among those who argue it is invalid are Nute [48], Bennett [3], van Rooij [51], Santorio [53], and Lassiter [38].

[6] Indeed, unlike with disjunctive antecedents, Starr [58] observes that *if or if*-conditionals *obligatorily* entail their simplifications, as shown by specificational disjunctive antecedent conditionals such as (i). (For discussion of specificational conditionals see Loewer [40], McKay and Inwagen [45], Nute [48], Bennett [3], and Klinedinst [34].)

(i) a. If John had taken the train or the metro, he would have taken the train.
b. #If John had taken the train or if he had taken the metro, he would have taken the train.

(9) If you had taken the train or if you had taken the metro, you would have been on time.

a. $\rightsquigarrow$ If you had taken the train, you would have been on time.
b. $\rightsquigarrow$ If you had taken the metro, you would have been on time.

Apart from Woods [65], Geurts [18] and Khoo [31], there has, however, been little discussion of disjunctions of whole conditionals—our focus in this paper.

2 Motivation from Theories of Wide Scope Free Choice

The puzzling behaviour of disjunctions of universal modals becomes especially important in light of recent interest in wide scope free choice, the inference from $\Diamond A \vee \Diamond B$ to $\Diamond A \wedge \Diamond B$, from Zimmermann [66], and illustrated in (10).

(10) a. He might be in Brixton or he might be in Victoria.
(i) $\rightsquigarrow$ He might be in Brixton.
(ii) $\rightsquigarrow$ He might be in Victoria.
b. You may go to Brixton or you may go to Victoria.
(i) $\rightsquigarrow$ You may go to Brixton.
(ii) $\rightsquigarrow$ You may go to Victoria.

In contrast, $\Box A \vee \Box B$ does not imply $\Box A \wedge \Box B$.

(11) a. He must be in Brixton or he must be in Victoria.
(i) $\not\rightsquigarrow$ He must be in Brixton.
(ii) $\not\rightsquigarrow$ He must be in Victoria.
b. He must go to Brixton or he must go to Victoria.
(i) $\not\rightsquigarrow$ He must go to Brixton.
(ii) $\not\rightsquigarrow$ He must go to Victoria.

A number of theories today, such as Zimmermann [66], Geurts [18], and Aloni [1], aim to account for wide scope free choice, the inference from $\Diamond A \vee \Diamond B$ to $\Diamond A \wedge \Diamond B$. It is important to check, however, that these theories do not inadvertently predict the analogous inference from $\Box A \vee \Box B$ to $\Box A \wedge \Box B$.

For example, Zimmermann [66] and Geurts [18] derive wide scope free choice by proposing that disjunctions denote conjunctions of possibilities: $A \vee B$ means $\Diamond A \wedge \Diamond B$, and thus $\Box A \vee \Box B$ is equivalent to $\Diamond \Box A \wedge \Diamond \Box B$. By a further principle—Zimmerman's Authority Principle, stating that the agent is an authority on what is permitted—this is equivalent to $\Box A \wedge \Box B$. This prediction, Geurts [18, 388] notes, "is clearly wrong".

Aloni [1] derives wide scope free choice assuming a very different semantics of disjunction, but along the way uses a constraint similar to the Authority Principle, which she calls *Indisputability*: all worlds in the speaker's epistemic state agree on which worlds are (deontically/epistemically/...) possible. If the speaker's accessibility relation is indisputable, Aloni predicts that $\Box A \vee \Box B$ implies $\Box A \wedge \Box B$ (we provide a proof of this fact in the Appendix). Without further refinement, then, Aloni's account risks making the same "clearly wrong" prediction as Zimmermann and Geurts.

When we examine disjunctions of universal modals, we see that one may explicitly affirm authority/indisputability without *Must A or must B* meaning *Must A and must B*.

(12) I am your parent, so I make the rules. And I'm telling you that you have to do your homework or you have to walk the dog.
$\not\rightsquigarrow$ You have to do your homework and you have to walk the dog.

The epistemic case in (1b) is also problematic, since it is natural to assume that one is an authority on one's epistemic state.

(1b) The keys must be in the drawer or John must have taken them.
$\not\rightsquigarrow$ The keys must be in the drawer and John must have taken them.

Do accounts of wide scope free choice such as Geurts' and Aloni's really predict the unwelcome inference from *Must A or must B* to *Must A and must B*? Or is there more to the interpretation of disjunctions of universal modals than meets the eye? In what follows we propose that there is: additional factors influence their interpretation which renders (12) unproblematic for accounts that derive a conjunctive reading of disjunctions of universal modals.

3 The Ingredients of Our Analysis

Our account has three, independently-motivated ingredients. Two are features of disjunction: its dynamic effect, and a conjunctive interpretation. The third relates the interpretation of modals to local contexts, what Mandelkern [42,43] calls *bounded modality*. We discuss each in turn.

3.1 The Dynamic Effect of Disjunction

The dynamic effect of disjunction is the fact that when we interpret a disjunction, we typically interpret the second disjunct assuming the negation of the first disjunct, and perhaps also symmetrically, interpret the first assuming the negation of the second. This is familiar from dynamic semantics (Heim [22], Veltman [61], Chierchia [9], and Beaver [2]), the wider literature on presupposition projection (Schlenker [54,55] and Chemla [8]), and what Klinedinst and Rothschild [33] call 'non-truth-tabular' disjunction.[7]

The primary evidence for disjunction's dynamic effect comes from presupposition projection. Karttunen [27] notes that both (13a) and its reordered variant (13b) do not presuppose that Jack has written letters.

[7] It is also similar to the contribution of *else*, though there are some differences between *or* and *or else* regarding which previous material can be negated, discussed by Weber et al. [63] and Meyer [46, 6–9].

(13) a. Either all of Jack's letters have been held up or he has not written any.
b. Either Jack has not written any letters or all of them have been held up.

The same point can be made with Evans' [12, 530] example in (14a), and Partee's bathroom sentences (appearing in Roberts [50]).[8]

(14) a. Either John does not own a donkey or he keeps it very quiet.
b. Either John keeps his donkey very quiet or he does not own a donkey.

(15) a. Either there's no bathroom in this house or it's in a funny place.
b. Either the bathroom is in a funny place or there's no bathroom here.

In these sentences, "all of Jack's letters have been held up" presupposes that there are such letters, "he keeps it very quiet" that John has a donkey, and "it's in a funny place" that the house has a bathroom. But the sentences as a whole do not: the presuppositions are filtered. A longstanding idea is that presuppositions must be satisfied in their local context (see Heim [23], Beaver [2], and Schlenker [55], among many others). Assuming that in a disjunction $A \vee B$, the local context of B entails $\neg A$, and the local context of A entails $\neg B$, the global context need not entail, respectively, that Jack has written letters, that John owns a donkey, and that the house has a bathroom. We therefore predict the correct presupposition filtering for these sentences.

A simple—but by no means the only—way to implement this idea is to introduce an additional parameter of interpretation representing the local context.[9]

$$[\![A \vee B]\!]^{w,c} = 1 \text{ iff } [\![A]\!]^{w,c} = 1 \quad \text{or } [\![B]\!]^{w,c\cap[\![\neg A]\!]^c} = 1 \qquad \text{ASYMMETRIC}$$

$$[\![A \vee B]\!]^{w,c} = 1 \text{ iff } [\![A]\!]^{w,c\cap[\![\neg B]\!]^c} = 1 \text{ or } [\![B]\!]^{w,c\cap[\![\neg A]\!]^c} = 1 \qquad \text{SYMMETRIC}$$

We will assume this additional parameter in what follows.

The pattern of presupposition projection remains when we add a universal modal to the second disjunct, as in (16), or both disjuncts, as in (17). We illustrate this here for Partee's bathroom sentences, though the same point applies to the other presupposition triggers considered above.

(16) a. There's no bathroom in this house or it must be in a funny place.
b. The bathroom is in a funny place or there must be no bathroom here.

(17) a. There must be no bathroom in this house or it must be in a funny place.

[8] Kalomoiros Schwarz [26] offer experimental evidence that presupposition filtering for disjunction is symmetric. For experimental evidence that presupposition filtering for conjunction is nonetheless asymmetric, see Mandelkern et al. [44].

[9] For any sentence A, $[\![A]\!]^c = \{w : [\![A]\!]^{w,c} = 1\}$ is the set of worlds where A is true.

b. The bathroom must be in a funny place or there must be no bathroom here.

In (16a), for example, we interpret "it must be in a funny place" assuming that the house has a bathroom. On the basis of these data, we propose that in $A \vee \Box B$ or $\Box A \vee \Box B$, the second disjunct may be interpreted assuming $\neg A$.[10]

Disjunctions of conditionals behave analogously. Given *If A, C, or if B, D*, we can interpret the second conditional assuming that A, the antecedent of the first, is false.

(18) a. If there's no bathroom in this house, I'll go home, or if it's in a funny place, I'll ask the host for directions.
b. If the bathroom is in a funny place, I'll ask the host for directions, or if there's no bathroom, I'll go home.

For example, in (18a) we interpret the second conditional assuming that the house has a bathroom.

Alternatively, given *If A, C, or if B, D*, we can interpret the second conditional assuming that C, the consequent of the first, is false.[11]

(19) a. If we hire architect A, our new office will have no bathroom, or if we hire architect B, it will be in a strange place.
b. If we hire architect B, our new office bathroom will be in a strange place, or if we hire architect A, our new office will have no bathroom.

In (19a), for example, we interpret the second conditional assuming that the new office has a bathroom. These provide evidence that when we interpret a disjunction of conditionals, the dynamic effect of disjunction can negate the antecedent or consequent of the first conditional.

Naturally, it remains possible for the disjunction's dynamic effect to negate the entire first conditional, as shown in (20) and (21).[12]

(20) The concrete stays intact if it is subjected to high pressure, or it is unsuitable for our purposes.
≡ The concrete stays intact if it is subjected to high pressure, or if it is not the case that it stays intact if it is subjected to high pressure, it is unsuitable for our purposes

(21) Ali will leave if Beth arrives, or they must have resolved their dispute.
a. ≡ Ali will leave if Beth arrives, or if it's not true that Ali will leave if Beth arrives, they must have resolved their dispute.

---

[10] It is an open and interesting question why it should be possible to negate the *prejacent* of a modal claim, rather than simply the whole modal claim. We discuss this in Sect. 6.

[11] I am grateful to Patrick Elliot, Alexandros Kalomoiros, Jacopo Romoli, and Yichi Zhang for helpful discussions of this example.

[12] (20) is modelled after the following example from Douven [10] of a left-nested conditional: "If this material becomes soft if it gets hot, it is not suited for our purposes.".

b. $\not\models$ Ali will leave if Beth arrives, or if Ali will not leave, they must have resolved their dispute.
c. $\not\models$ Ali will leave if Beth arrives, or if Beth will not arrive, they must have resolved their dispute.

Negating the entire first disjunct is not very interesting for present purposes since it has the same truth conditions as when the dynamic effect does not apply. One way to see this is to note that $A \vee B$, $A \vee (\neg A \to B)$, and $(\neg B \to A) \vee (\neg A \to B)$ are all classically equivalent, where $\to$ denotes the material conditional (that is, $A \to B$ is equivalent to $\neg A \vee B$).[13]

3.2 Conjunctive Interpretations of Disjunction

The second ingredient of our proposal is that disjunction may receive a conjunctive interpretation, familiar from the literature on wide scope free choice (Zimmermann [66], Geurts [18], and Aloni [1]), already encountered in Sect. 2, and Klinedinst and Rothschild's [33] non-truth-tabular disjunction. In what follows we remain neutral on what exact mechanism is responsible for wide scope free choice; for example, Aloni's [1] state-based account or Goldstein's [20] homogeneous dynamic semantics. Provided we have some mechanism to strengthen disjunctions into conjunctions, one may add it as a module to our account.

3.3 Bounded Modality

For our account to succeed, we need to relate the interpretation of modal statements to local contexts. This is the third and final ingredient of our proposal: the interpretation of modal statements is restricted to local contexts. Mandelkern [42,43] independently argues for this constraint, calling it *bounded modality.*

On a simple semantics of modals via accessibility relations, we can implement the constraint by requiring that modals are evaluated at the accessible worlds compatible with the local context, so that $\Box A$ is true just in case A is true at every accessible world compatible with the local context. Formally, let W to be the set of possible worlds, R the relevant accessibility relation for modals, and $R[w] = \{w' \in W : wRw'\}$ the set of accessible worlds at w.[14] *Must* A is true at

[13] Granted, the material conditional is likely not the most natural restriction device by which to represent the effect of local contexts. Nonetheless, however one represents the restriction provided by local contexts, it is plausible that disjunctions will have the same truth conditions regardless of whether or not the local context for the second disjunct entails the negation of the first, and whether or not the local context for the first entails the negation of the second. Indeed, precisely this assumption—that the addition of local contexts preserves truth conditions—is a cornerstone of the most comprehensive theory of local contexts currently available: Schlenker's [55] algorithm.

[14] We assume that for each modal there is an associated accessibility relation; to distinguish them one may choose to add indices to modals and accessibility relations, *à la* von Fintel [14].

a world w just in case A is true at every world compatible with the local context that is accessible from w.[15]

(22) **Bounded Kripke.**
$[\![\Box A]\!]^{c,w}$ is true just in case $R[w] \cap c \subseteq [\![A]\!]^c$.

On a more sophisticated story, such as Kratzer's [36] doubly-relative account, we implement boundedness by restricting the modal base to the local context. Kratzer [35] proposes that *Must A* holds just in case A is true at every world in the modal base ranked highest according to the ordering source: $[\![\Box A]\!]^{w,f,g,c}$ is true just in case $\max_{g(w)} \bigcap f(w) \subseteq [\![A]\!]^{f,g,c}$.[16] On the bounded version, *Must A* is true at a world w just in case A is true at every world in the modal base at w *compatible with the local context* that is ranked highest according to the ordering source at w.

(23) **Bounded Kratzer.**
$[\![\Box A]\!]^{f,g,c,w}$ is true just in case $\max_{g(w)}((\bigcap f(w)) \cap c) \subseteq [\![A]\!]^{f,g,c}$.

We can derive bounded modality from independent observations. For one may think of the local context as itself a kind of presupposition. At every stage of interpretation we presuppose that the actual world is compatible with the local context. The local context is, if you like, the arch-presupposition: the presupposition that entails all others at each stage of interpretation.

Heim [24] notes that, when we interpret a modal statement, we assume that its presuppositions hold throughout the modal base.

(24) a. Patrick wants to sell his cello.
b. If John had attended the party too, ...

As Heim observes, (24a) presupposes that in all of the worlds compatible with Partick's beliefs, he owns a cello. It is assertable, say, even if Patrick is only under the misconception that he owns a cello. Similarly, when it is already in the

---

15 There are two noteworthy differences between our entry in (22) and the bounded theory of Mandelkern [42,43]. For Mandelkern, modal statements *presuppose* that the modal domain is included in the local context, whereas we implement this restriction via the *truth conditions* of modal statements. Our simplification is purely for reasons of simplicity: when showing results for our system, we can assume a uniform modal base, without needing to compare various modal bases under different restrictions. One may straightforwardly implement our account on a more nuanced story, such as Mandelkern's, which distinguishes between truth and definedness conditions. The second difference is that we allow the local context to restrict the prejacent of a modal, while Mandelkern does not consider this possibility.

16 We let $[\![A]\!]^{f,g,c} = \{w' : [\![A]\!]^{w',f,g,c} = 1\}$ denote the set of worlds where A is true, and $[\![\neg A]\!]^{f,g,c} = W \setminus \{w' : [\![A]\!]^{w',f,g,c} = 1\}$ the set of worlds where A is false, with respect to f, g, and c. For every set of worlds $X \subseteq W$, we let $\max_P(X) = \{w \in X : w' <_P w \text{ for no } w' \in X\}$, where for any set of propositions P, $w <_P w'$ holds just in case for all $p \in P$ such that p is true at w', p is true at w (for discussion see von Fintel and Heim [16]).

common ground that Mary attended, when we interpret (24b) we only consider cases where Mary still attended. We could not assert, for example, *If John had attended the party too, Mary would not have attended*, when Mary is among the salient individuals picked out by *too*.

It is typically assumed that bouletic predicates such as *want* have a doxastic modal base: they are restricted to worlds compatible with the attitude holder's beliefs (Heim [24] and von Fintel [15]). Then (24a) shows that the presuppositions of a modal statement must be satisfied in the modal base. (24b) makes the same point with a counterfactual.

So we have two observations. The local context is itself a presupposition, and modal statements are evaluated at modal bases restricted to worlds where the statement's presuppositions hold. With these we derive bounded modality.

4 Disjunctions of Universal Modals

4.1 Putting the Ingredients Together

Putting the above ingredients together, on our proposal the meaning of disjunctions of universal modals *Must A or must B* as in (1) can be paraphrased as follows, with an asymmetric and symmetric dynamic effect, respectively.

(25) ASYMMETRIC DYNAMIC EFFECT + CONJUNCTIVE INTERPRETATION
- a. You must do clean your room and if you do not clean your room, you must walk the dog.
- b. The keys must be in the drawer and if they are not in the drawer, John must have taken them.

(26) SYMMETRIC DYNAMIC EFFECT + CONJUNCTIVE INTERPRETATION
- a. If you do not walk the dog, you must do clean your room, and if you do not clean your room, you must walk the dog.
- b. If John has not taken the keys, they must be in the drawer, and if they are not in the drawer, John must have taken them.

We propose that both the asymmetric and symmetric readings are available, with the asymmetric reading giving priority to the first disjunct and the second true at more remote possibilities—what Schwager [56] calls the *or else* effect. For example, the asymmetric reading allows us to predict the following contrast.

(27) a. You have to pay the bill, or you have to go to jail.
b. ??You have to go to jail, or you have to pay the bill.

Given an asymmetric dynamic interpretation and conjunctive reading, these wind up with a meaning we may paraphrase as follows, which pattern the same as (27) in their acceptability.

(28) a. You have to pay the bill. If you do not pay the bill, you have to go to jail.

b. ?? You have to go to jail. If you do not go to jail, you have to pay the bill.

Following Meyer [46], we adopt Kratzer's [36] analysis of modality, whereby worlds are ordered along a dimension according to their flavour (such as epistemic, or in this case, deontic). $\Box A$ is true at a world w just in case A is true at all worlds in the modal base that come closest to the ideal according to w. In addition, the contribution of *if*-clauses is to restrict the modal base (Kratzer [37]). Thus (28a) says that at all of the normatively best worlds, you pay the bill, and at all of the normatively best worlds where you don't, you go to jail.

In this way sentences like (28a) are of a piece with well-known counterexamples to antecedent strengthening for conditionals, such as Goodman's [21] classic example: we can accept *If I struck this match, it would light* while denying *If I struck this match and it were wet, it would light.* (28a) is true while (28b) is false since all else being equal, worlds where you pay the bill are better—closer to our deontic ideals—than worlds where you go to jail.

In contrast, on the symmetric reading both options are viewed on a par, with neither having priority over the other. This is shown in examples like (29), where order seems not to matter.

(29) *For Ali to win the game, the die must land on multiple of three.*
a. Ali has to roll a three or he has to roll a six.
b. Ali has to roll a six or he has to roll a three.

Given a symmetric dynamic interpretation and conjunctive reading of conjunction, these can be paraphrased as follows.

(30) a. If Ali doesn't roll a six, he has to roll a three, and if he doesn't roll a three, he has to roll a six.
b. If Ali doesn't roll a three, he has to roll a six, and if he doesn't roll a six, he has to roll a three.

This proposal solves the problem on behalf of Zimmermann's, Geurts' and Aloni's theories of wide scope free choice. Instead of trying to block the unwanted conjunctive reading of $\Box A \vee \Box B$ as $\Box A \wedge \Box B$, we embrace it. While the conjunctive reading initially appeared too strong, disjunction's dynamic effect weakens it, resulting in plausible predictions for what disjunctions of universal modals intuitively mean.

4.2 The Equivalence of *Must A or Must B* and *Must*(*A or B*)

Given a symmetric dynamic effect we can derive the equivalence of $\Box A \vee \Box B$ and $\Box(A \vee B)$. Generally speaking, the key observation is that $d \cap \overline{a} \subseteq b$ is equivalent to $d \subseteq a \cup b$ for any sets a, b and d (where $\overline{a}$ denotes a's complement). Taking d to be the modal domain, a the proposition expressed by A and b that of B, we derive the equivalence of $\Box A \vee \Box B$ and $\Box(A \vee B)$.

Let us first show this on the bounded Kripkean semantics. As a convenient shorthand, let us write $\Box_A B$ to denote that $\Box B$ is evaluated at the global context restricted by A.

$$[\![\Box_A B]\!]^{w,f,g,c} \quad = \quad [\![\Box B]\!]^{w,f,g,c\cap[\![A]\!]^{f,g,c}}$$

It follows that $\Box_{\neg A} B$ is equivalent to $\Box(\neg A \rightarrow B)$, where $\rightarrow$ is the material conditional, which is equivalent to $\Box(A \vee B)$. Hence $\Box_{\neg B} A \wedge \Box_{\neg A} B$ is equivalent to $\Box(A \vee B)$.

$$\begin{array}{ll} & \Box_{\neg B} A \wedge \Box_{\neg A} B \\ \equiv & \Box(\neg B \rightarrow A) \wedge \Box(\neg A \rightarrow B) \\ \equiv & \Box(A \vee B) \wedge \Box(A \vee B) \\ \equiv & \Box(A \vee B) \end{array}$$

To show this equivalence on the bounded Kratzerian semantics, we need two further assumptions. The first is that $\neg A$ and $\neg B$ are each possible: $\neg\Box A$ and $\neg\Box B$. The second is a principle we will call *Strengthening with a Possibility*, named after a corresponding rule in the logic of conditionals: given $\neg$(*if* $A, \neg B$), *if* A, C implies *if* A *and* B, C.

Strengthening with a Possibility (SP).
If $(\max_P X) \cap Y$ is nonempty, then $\max_P(X \cap Y) = (\max_P X) \cap Y$.

Strengthening with a Possibility says that if we find a world where Y is true among the closest X-worlds, then the closest X-and-Y-worlds are just the closest X-worlds where Y is true. This does not hold in general on ordering approaches to modality. It corresponds to *almost connectedness*: for any worlds w, w', w'', if $w \leq_P w''$ then either $w \leq_P w'$ or $w' \leq_P w''$ (Veltman [60, 103]). It is automatically satisfied by the sphere models of Lewis [39], though there are—to my mind convincing—counterexamples to it (Ginsberg [19, 50], Boylan and Schultheis [4,5]). Nonetheless, in the simple kinds of modal statements we have considered, we can reasonably expect it to hold.

Assuming the bounded Kratzerian semantics in (23), Strengthening with a Possibility guarantees that if there is a world where A is true among the closest worlds, then $\Box_A B$ is equivalent to $\Box(A \rightarrow B)$.[17]

$$SP \wedge \Diamond A \quad \rightarrow \quad (\Box_A B \leftrightarrow \Box(A \rightarrow B))$$

[17] Let us prove this here. Pick any parameters w, f, g, c. To simplify notation, for any sentence A let $|A| = [\![A]\!]^{f,g,c}$ be the set of worlds where A is true, $D = \max_{g(w)}((\bigcap f(w)) \cap c)$, the modal domain, that is, the closest worlds in the modal base restricted to the global context, and let $D_A = \max_{g(w)}((\bigcap f(w)) \cap c \cap |A|)$ be the modal domain where the local context is restricted by A. Suppose $\Diamond A$ is true at w. Then $D \cap |A|$ is nonempty, so by Strengthening with a Possibility, $D_A = D \cap |A|$. Hence $\Box_A B$ is true at w iff $D_A \subseteq |B|$ iff $D \cap |A| \subseteq |B|$ iff $D \subseteq |\neg A| \cup |B|$ iff $\Box(A \rightarrow B)$ is true at w.

Hence if $\neg A$ is possible, $\Box_{\neg A} B$ is equivalent to $\Box(\neg A \rightarrow B)$, which is equivalent to $\Box(A \vee B)$, and if $\neg B$ is possible, $\Box_{\neg B} A$ is also equivalent to $\Box(A \vee B)$. Hence at any world where $\neg A$ and $\neg B$ are possible, $\Box_{\neg B} A \wedge \Box_{\neg A} B$ is equivalent to $\Box(A \vee B)$.

$$SP \wedge \Diamond\neg A \wedge \Diamond\neg B \quad \rightarrow \quad ((\Box_{\neg B} A \wedge \Box_{\neg A} B) \leftrightarrow \Box(A \vee B))$$

Returning to our original examples in (1), this accounts for the following equivalences.

(31) a. You must do clean your room or you must walk the dog.
b. You must do clean your room or walk the dog.

(32) a. The keys must be in the drawer or John must have taken them.
b. It must be that the keys are in the drawer or John took them.

Happily, our account does not inadvertently predict this equivalence for universal quantifiers in general, such as in (3), repeated below.

(33) a. Everyone is in the kitchen or the garden.
b. Everyone is in the kitchen or everyone is in the garden.

A local context is a piece of information—in possible worlds semantics, a set of worlds—rather than a set of individuals. A set of individuals (say, the set of salient people outside the kitchen) is simply not the kind of thing that can serve as a local context.

5 Disjunctions of Conditionals

Turning to conditionals, the very same combination of disjunction's dynamic effect and conjunctive interpretation accounts for the contrast in (2), repeated below, where the first appears to be interpreted conjunctively and the second not.

(2) a. If Alice had come to the party, Charlie would have come. Or if Bob had come, Charlie would have come.
b. If Alice had come to the party, Charlie would have come. Or if Alice had come, Darius would have come.

In principle there are a number of options for which clause is negated by the dynamic effect of disjunction: the whole conditional, the antecedent or the consequent. As discussed, negating the whole conditional is possible, but not very interesting for our purposes since it is classically equivalent to an interpretation without any dynamic effect.

Putting aside the option of negating the whole conditional, we find that in (2a), negating the previous consequent is not an option since it would violate a general ban on triviality; the second conditional would assert that if Bob but not Charlie had come, Charlie would have come. If there are relevant possibilities where Bob but not Charlie comes, this is trivially false. If there are not, the conditional either suffers from presupposition failure or is vacuously true, depending on one's view. Either way, the conditional is uninformative, violating principles of conversation. However, negating the previous antecedent is perfectly possible, stating that if Alice but not Bob had come to the party, Charlie would have.

(34) If A, C or if B, C

a. If A, C and if B and $\neg C$, C NEGATING THE CONSEQUENT: ✗

b. If A, C and if B and $\neg A$, C NEGATING THE ANTECEDENT: ✓

In (2b) the situation is reversed. Negating the antecedent is not an option since it would also violate a general ban on triviality; the second conditional would assert, vacuously, that if Alice but not Alice had come, Darius would have come. However, negating the previous consequent is an option, stating that if Alice but not Charlie had come, Darius would have.

(35) If A, C or if A, D

a. If A, C and if A and $\neg C$, D NEGATING THE CONSEQUENT: ✓

b. If A, C and if A and $\neg A$, D NEGATING THE ANTECEDENT: ✗

Given an asymmetric dynamic effect, then, the sentences in (2) have interpretations which we may paraphrase as follows.

(36) ASYMMETRIC DYNAMIC EFFECT + CONJUNCTIVE INTERPRETATION

a. If Alice had come to the party, Charlie would have come. And if Bob but not Alice had come, Charlie would have come.

b. If Alice had come to the party, Charlie would have come. And if Alice but not Charlie had come, Darius would have come.

And given a symmetric dynamic effect, they have interpretations which we may paraphrase as:

(37) SYMMETRIC DYNAMIC EFFECT + CONJUNCTIVE INTERPRETATION

a. If Alice but not Bob had come to the party, Charlie would have come. And if Bob but not Alice had come, Charlie would have come.

b. If Alice but not Darius had come to the party, Charlie would have come. And if Alice but not Charlie had come, Darius would have come.

These are plausible predictions for what (2) mean.

This also correctly predicts the interpretation of (6), where the antecedent and consequents are distinct. For example, on a symmetric dynamic effect on the antecedent, the meaning of (6) can be paraphrased as in (38).

(38) If you had taken the morning train (and not taken the evening train), you would have arrived before lunch. And if you had taken the evening

train (and not taken the morning train), you would have arrived after lunch.

This solves our puzzle regarding why (2a) appeared to be interpreted conjunctively but (2b) disjunctively. In fact both are read conjunctively. The restriction provided by local contexts results in the apparent divergence in readings. This has the welcome consquence of dissolving the tricky question of explaining why the conjunctive inference applies when the antecedents are the same but disappears when the consequents are the same. In fact the inference applies across the board.

5.1 The Equivalence of *if A, C or if A, D* and *if A, C or D*

Following Kratzer [37], we assume the conditionals feature a modal (overt or covert), and that conditional antecedents restrict the modal base.

$$[\![\textit{if } A, C]\!]^{w,f,g,c} \quad = \quad [\![C]\!]^{w,f+A,g,c}$$

where $f + A$ is given by $f + A(w) = f(w) \cup \{[\![A]\!]^{f,g,c}\}$ for every world w.

Note that on our bounded semantics of modals in (23), we could equivalently write that conditional antecedents restrict the local context: *if* $A, \Box C$ is equivalent to $\Box_A C$. Then assuming a symmetric dynamic effect of disjunction and a conjunctive interpretation, *if A, C or if A, D* expresses $\Box_{A\wedge\neg D}C \wedge \Box_{A\wedge\neg C}D$. Under the same assumptions as before—Strengthening with a Possibility and that $\Diamond_A\neg C$ and $\Diamond_A\neg D$ are both true—we predict $\Box_{A\wedge\neg D}C$ to be equivalent to $\Box_A(\neg D \rightarrow C)$, and hence to $\Box_A(C \vee D)$; similarly, we predict $\Box_{A\wedge\neg C}D$ to be equivalent to $\Box_A(C\vee D)$. Then their conjunction is also equivalent to $\Box_A(C\vee D)$.

$$SP \wedge \Diamond_A\neg C \wedge \Diamond_A\neg D \quad \rightarrow \quad ((\Box_{A\wedge\neg D}C \wedge \Box_{A\wedge\neg C}D) \leftrightarrow \Box_A(C \vee D))$$

Thus assuming Strengthening with a Possibility, $\Diamond_A\neg C$ and $\Diamond_A\neg D$, we predict *if A, C or if A, D* to be equivalent to *if A, C or D*. This prediction is borne out.

(39) a. If Alice had come to the party, Charlie would have come, or if Alice had come to the party, Darius would have come.
b. If Alice had come to the party, Charlie or Darius would have come.

Importantly, we derive this equivalence assuming that disjunctions of conditionals receive a conjunctive interpretation across the board. This resolves the puzzle of explaining why *If A, C or if B, C* seems to receive a conjunctive interpretation while *If A, C or if A, D* seems not to. In fact both are read conjunctively, but the symmetric dynamic effect of disjunction weakens them, rendering *If A, C or if A, D* equivalent to *If A, C or D*.

A further test of this prediction comes from (40), modelled after Quine's [49] Bizet–Verdi example.

(40) *Coin A landed heads, coin B landed tails.*

a. If both coins had landed on the same side, they both would have landed heads, or if they had landed on the same side, they both would have landed tails.
b. If both coins had landed on the same side, they would have landed both heads or both tails.

These are naturally felt to be equivalent, as predicted on this account.[18]

6 Negating the Entire Disjunct or a Part Thereof

We have proposed that the prejacent of a modal statement is available for restriction by local contexts: when we interpret $\Box A \vee \Box B$, we may evaluate each disjunct assuming that the other disjunct is false, or assuming that *prejacent* of the other disjunct (that is, A and B) is false. Furthermore, we proposed that in a conditional, the antecedent and consequent are both available for restriction. This is not predicted by standard approaches to local contexts, such as Schlenker's [55] algorithm, but there is independent evidence that these restrictions are available. In Sect. 3.1 we have seen independent evidence for this assumption from presupposition projection. Furthermore, Meyer [46] observes the same behaviour for *or else* sentences with modals, offering examples such as (41), found in Heim and Kratzer [25].

(41) Pronouns must be generated with an index or else they will be uninterpretatable.

As Meyer points out, the second disjunct is interpreted assuming that pronouns are not generated with an index, rather than that they merely need not be generated with an index.

The same observation applies to plain disjunctions without *else*, as in (42a), which receives the same interpretation as (41).

(42) a. Pronouns must be generated with an index or they will be uninterpretatable.

[18] As with all cases of wide-scope free choice, the first also has an ignorance reading—brought out by the continuation ... *but I don't know which*—on which they are not equivalent. We can express this reading on the present proposal as $\Box A \vee \Box B$, without the addition of local contexts, or with local contexts where the local context is the negation of the entire other disjunct: asymmetrically as $\Box A \vee \Box_{\neg\Box A} B$, and symmetrically as $\Box_{\neg\Box B} A \vee \Box_{\neg\Box A} B$.

Note further that the equivalence of *If A, C or if A, D* and *If A, C or D* is also predicted by selectional approaches to conditionals, which evaluate the consequent at a unique selected world where the antecedent holds (see, for example, Stalnaker [57], Cariani and Santorio [7], Cariani [6], and Mandelkern [41]). We therefore cannot take the intuitive equivalence of (40a) and (40b) as a decisive point in favour of our proposal, though it is nonetheless a welcome result that the correct prediction also follows on our proposal.

Example (43), from the film *The Terminator*, illustrates the point with a naturally-occurring example. John comes back from the future to tell Sarah:

(43) You must survive or I will never exist.

This says that if Sarah does not survive, John will never exist, rather than that if Sarah is not required to survive, John will never exist. This suggests that both the modal statement as a whole and its prejacent are in general accessible for restriction by local contexts.

Plain modal statements appear to be special in this respect. Compare:

(44) a. You must clean your room or you must walk the dog.
b. You are required to clean your room or you are required to walk the dog.
c. According to the rules, you must clean your room, or according to the rules, you must walk the dog.

We typically read (44a) as equivalent to *You must clean your room or walk the dog.* In contrast, (44b) and (44c) seem to be more naturally read as saying that one of the disjuncts is required, but the speaker is unsure which.[19]

We see the same with epistemic modals.

(45) a. The keys must be in the drawer or John must have taken them.
b. I'm certain that the keys are in the drawer or I'm certain that John has taken them.
c. According to my information, the keys must be in the drawer, or according to my information, John must have taken them.

(45a) has an easily accessible reading on which it is equivalent to $\Box(A \vee B)$, but this reading appears harder to access in (45b) and (45c).

One possible, simple explanation for these differences comes from pragmatic pressure to avoid redundancy. By adding more material, such that *according to the rules* or *it is required that*, it becomes more costly to assert $\Box A \vee \Box B$ rather than $\Box(A \vee B)$. With greater cost, we are more likely to assume that there must be a good reason for this additional cost, and so assume that the speaker does not intend to communicate the simpler $\Box(A \vee B)$. As we have seen, $\Box_{\neg B} A \vee \Box_{\neg A} B$ is equivalent to $\Box(A \vee B)$. However, the alternative available parse, $\Box_{\neg \Box B} A \vee \Box_{\neg \Box A} B$ is not equivalent to $\Box(A \vee B)$, with greater cost we would therefore be more likely to opt for the second reading. I will leave this as a speculative idea for now.

Zooming out, there is also evidence that given a disjunction $A \vee B$, the local context for B cannot be restricted to just any part of A whatsoever. Meyer

[19] The empirical picture here is subtle. I do not rule out that (44b) and (44c) may have a reading equivalent to $\Box(A \vee B)$, further empirical testing would prove beneficial.

[46] presents the following argument against this idea, based on presupposition projection.[20]

(46) Either they didn't remind John to bring his ID or Bill regretted that he didn't (bring it).
Presupposes: *John didn't bring his ID* (Meyer [46], example 22)

If we assume that *John brought his ID* is a subpart of *They didn't remind John to bring his ID*, then we might expect that the second disjunction could be interpreted assuming that he didn't bring his ID, in which case the presupposition that John didn't bring his ID would incorrectly predicted to be filtered.

Thankfully, the present proposal does not require that any part of a disjunct is available for restriction, but merely the more limited claim that in a plain modal statement, *Must A or must B*, the prejacent of each modal is available for restriction, and that in a conditional, the antecedent and consequent are both available for restriction. At present we lack a general theory of what parts of a statement are accessible for restriction by local contexts, though the data in (44) and (45) help point us toward an answer.

7 Conclusion

Disjunctions of universal modals and disjunctions of conditionals give rise to surprising effects, ones unexpected from the perspective of classical logic. We often read *Must A or must B* as equivalent to *Must(A or B)*. And while *if A, C or if B, C* intuitively implies each conditional, *if A, C or if A, D* does not.

We have proposed a uniform account of these data. Our account combined three independently motivated ingredients: the dynamic effect of disjunction (familiar from presupposition projection), a conjunctive interpretation (familiar from wide scope free choice), and bounded modality (familiar from Mandelkern's [42] account of *might*). As we have seen, these three ingredients together with some mild auxiliary assumptions, we derive the equivalence of *Must A or must*

[20] An anonymous reviewer helpfully points out that the same point applies to conjunction:

(i) Mary believes that Bill will come, and Sue knows that Bill will come.

It is standardly assumed that the local context for B in A *and* B entails A. If it were possible for local context to entail any subpart of A whatsoever, then the local context for the second conjunct could entail that Bill will come (rather than the full conjunct, Mary believes that Bill will come). This would incorrectly predict presupposition that Bill will come to be filtered.

It appears that an analogous point can be made for disjunction.

(ii) Mary believes that Bill will come, or Sue knows that he won't.

This seems to presuppose that Bill will not come, though the judgement is not perfectly robust and would benefit from further experimental investigation.

B and *Must*(A *or* B), and explain why *if A, C or if B, C* is read conjunctively, while *if A, C or if A, D* appears to be read disjunctively. In fact, both are read conjunctively, but the dynamic effect of disjunction weakens the latter, rendering it equivalent to *if A, C or D.*

Acknowledgments. For fruitful discussion of the present material, I very am grateful to Arseny Anisimov, Patrick Elliot, Alexandros Kalomoiros, Matt Mandelkern, Jacopo Romoli, Yichi Zhang, the participants of The Fourth Tsinghua Interdisciplinary Workshop on Logic, Language, and Meaning (TLLM IV) and the semantics seminar at Heinrich Heine University Düsseldorf.

Disclosure of Interests. The authors have no competing interests to declare that are relevant to the content of this article.

Appendix

The Inference from $\Box A \vee \Box B$ to $\Box A \wedge \Box B$ in Aloni's (2023) System

Let $M = (W, R, V)$ be a Kripke model and $s \subseteq W$ a set of worlds. For any world $w \in W$ let $R[w] = \{w' : wRw'\}$ be the set of worlds accessible from w. Aloni introduces a constraint (called 'neglect-zero') requiring each state to be nonempty, represented by a constant NE. She proposes that sentences are interpreted by a process of pragmatic enrichment that avoids empty interpretations; for example, a disjunction $A \vee B$ is interpreted as $(A \wedge \text{NE}) \vee (B \wedge \text{NE})$.

$$
\begin{array}{rll}
M, s \models A \vee B & \text{iff} & \text{there are states } t \text{ and } t' \text{ such that } s = t \cup t', \\
 & & M, t \models A \text{ and } M, t' \models B \\
M, s \models A \wedge B & \text{iff} & M, s \models A \text{ and } M, s \models B \\
M, s \models \Box A & \text{iff} & \text{for all } w \in s, M, R[w] \models A \\
M, s \models \text{NE} & \text{iff} & s \neq \emptyset
\end{array}
$$

We call R *indisputable* in (M, s) just in case for all $w, v \in s$, $R[w] = R[v]$.

Given these semantic clauses and the assumption that R is indisputable in (M, s), it follows that disjunctions of universal modals imply their conjunctions: $(\Box A \wedge \text{NE}) \vee (\Box B \wedge \text{NE})$ implies $(\Box A \wedge \text{NE}) \wedge (\Box B \wedge \text{NE})$. For $M, s \models (\Box A \wedge \text{NE}) \vee (\Box B \wedge \text{NE})$ holds just in case there are non-empty states t and t' such that $s = t \cup t'$, (i) for all $w \in t$, $M, R[w] \models A$ and (ii) for all $w' \in t'$, $M, R[w'] \models B$. Since t and t' are nonempty, there are $w \in t$ and $w' \in t'$, and by indisputability, $R[w] = R[w']$. To show that $M, s \models \Box A \wedge \Box B$, pick any $v \in s$. By indisputability, $R[v] = R[w] = R[w']$. Then $M, R[v] \models A$ and $M, R[v] \models B$, so $M, s \models \Box A \wedge \Box B$, and since s is nonempty, also $M, s \models (\Box A \wedge \text{NE}) \wedge (\Box B \wedge \text{NE})$.

Links to Online Materials

- Mandarin
  Chinese *huò* https://web.archive.org/web/20220425194146/https://www.biblegateway.com/passage/?search=Leviticus+5&version=CNVT

- English *or* http://web.archive.org/web/20220408125727/https://www.biblegateway.com/passage/?search=Leviticus+5&version=KJV
- Hebrew *o* https://web.archive.org/web/20220425194625/https://www.biblegateway.com/passage/?search=Leviticus+5&version=WLC
- Hungarian *vagy* https://web.archive.org/web/20220425194655/https://www.biblegateway.com/passage/?search=Leviticus+5&version=KAR
- Icelandic *eða* https://web.archive.org/web/20220425194740/https://www.biblegateway.com/passage/?search=Leviticus+5&version=ICELAND
- Māori *rānei* http://web.archive.org/web/20201127071407/https://www.biblegateway.com/passage/?search=Leviticus+5&version=MAORI
- Urdu *yâ* https://web.archive.org/web/20220425195430/https://www.biblegateway.com/passage/?search=Leviticus+5&version=ERV-UR
- Somali *ama* https://web.archive.org/web/20220425195432/https://www.biblegateway.com/passage/?search=Leviticus+5&version=SOM
- Welsh *neu* http://web.archive.org/web/20220425195453/https://www.biblegateway.com/passage/?search=Leviticus+5&version=BWM
- Yoruba *tàbí* https://web.archive.org/web/20220425195612/https://www.biblegateway.com/passage/?search=Leviticus+5&version=BYO
- Example (43) https://web.archive.org/web/20241222204426/https://getyarn.io/yarn-clip/dfd80466-cd89-421b-b3df-9201282cd90f

References

1. Aloni, M.: Logic and conversation: The case of free choice. Semantics and Pragmatics 15, May 2023. https://doi.org/10.3765/sp.15.5.
2. Beaver, D.: Presupposition and assertion in dynamic semantics. vol. 29. CSLI publications (2001)
3. Bennett, J.: A philosophical guide to conditionals. Oxford University Press (2003)
4. Boylan, D., Schultheis, G.: Strengthening principles and counterfactual semantics. In: Proceedings of the 21st Amsterdam Colloquium (2017). Ed. by Floris Roelofsen Alexandre Cremers Thom van Gessel, pp. 155–164. https://semanticsarchive.net/Archive/jZiM2FhZ/AC2017-Proceedings.pdf
5. Boylan, D., Schultheis, G.: How strong is a counterfactual? J. Philosophy **118**(7), 373–404 (2021). https://doi.org/10.5840/jphil2021118728
6. Cariani, F.: The Modal Future: A Theory of Future-Directed Thought and Talk. Cambridge University Press (2021)
7. Cariani, F., Santorio, P.: Will done better: selection semantics, future credence, and indeterminacy. Mind **127**(505), 129–165 (2018). https://doi.org/10.1093/mind/fzw004
8. Chemla, E.: Similarity: Towards a unified account of scalar implicatures, free choice permission and presupposition projection. Under revision for Semantics and Pragmatics (2009)
9. Chierchia, G.: Dynamics of Meaning: Anaphora, Presupposition and the Theory of Grammar. University of Chicago Press (1995)
10. Douven, I.: On de Finetti on Iterated Conditionals. Computational models of rationality: Essays Dedicated to Gabriele Kern-Isberner on the Occasion of Her 60th Birthday, pp. 265–279 (2016)

11. Ellis, B., Jackson, F., Pargetter, R.: An objection to possibleworld semantics for counterfactual logics. J. Philos. Log. **6**(1), 355–357 (1977). https://doi.org/10.1007/BF00262069
12. Evans, G.: Pronouns, quantifiers, and relative clauses (1). Can. J. Philos. **7**(3), 467–536 (1977). https://doi.org/10.1080/00455091.1977.10717030
13. Fine, K.: Counterfactuals without possible worlds. J. Philosophy **109**(3), 221–246 (2012). https://doi.org/10.5840/jphil201210938
14. von Fintel, K.: Restrictions on quantifier domains. PhD thesis. University of Massachusetts at Amherst (1994). https://semanticsarchive.net/Archive/jA3N2IwN/fintel-1994-thesis.pdf
15. von Fintel, K.: NPI licensing, Strawson entailment, and context dependency. J. Semant. **16**(2), 97–148 (1999). https://doi.org/10.1093/jos/16.2.97
16. von Fintel, K., Heim, I.: Intensional semantics. MIT Textbook (2011). http://web.mit.edu/fintel/fintel-heim-intensional.pdf
17. von Fintel, K., Iatridou, S.: A modest proposal for the meaning of imperatives. Modality across syntactic categories, pp. 288–319 (2017). http://web.mit.edu/fintel/fintel-iatridou-2015-modest.pdf
18. Geurts, B.: Entertaining alternatives: disjunctions as modals. Nat. Lang. Seman. **13**(4), 383–410 (2005). https://doi.org/10.1007/s11050-005-2052-4
19. Matthew, L.: Ginsberg. Counterfactuals. Artif. Intell. **30**, 35–79 (1986). https://doi.org/10.1016/0004-3702(86)90067-6
20. Goldstein, S.: Free choice and homogeneity. Semantics and Pragmatics 12 (23 2019), pp. 1–47. https://doi.org/10.3765/sp.12.23.
21. Goodman, N.: The Problem of Counterfactual Conditionals. J. Philosophy **44**(5), 113–128 (1947). https://doi.org/10.2307/2019988
22. Heim, I.: The semantics of definite and indefinite noun phrases. PhD thesis. University of Massachusetts, Amherst (1982)
23. Heim, I.: On the projection problem for presuppositions. Second West Coast Conference on Formal Linguistics. Ed. by D. Flickinger. Stanford University Press, pp. 114–125 (1983)
24. Heim, I.: Presupposition projection and the semantics of attitude verbs. J. Semant. **9**(3), 183–221 (1992). https://doi.org/10.1093/jos/9.3.183
25. Heim, I., Kratzer, A.: Semantics in generative grammar. Blackwell Oxford (1998)
26. Kalomoiros, A., Schwarz, F.: Presupposition projection from 'and' vs 'or': Experimental data and theoretical implications. J. Semantics (2024)
27. Karttunen, L.: Presuppositions of compound sentences. Linguistic inquiry **4**(2), 169–193 (1973)
28. Kaufmann, M., Whitman, J.: Conditional conjunctions informed by Japanese and Korean. Linguistics Vanguard **8**(s4), 479–490 (2022)
29. Keshet, E., Medeiros, D.J.: Imperatives under coordination. Natural Lang. Linguistic Theory **37**(3), 869–914 (2018). https://doi.org/10.1007/s11049-018-9427-y
30. Khoo, J.: Coordinating Ifs. J. Semant. **38**(2), 341–361 (2021). https://doi.org/10.1093/jos/ffab006
31. Khoo, J.: Disjunctive antecedent conditionals. Synthese **198**(8), 7401–7430 (2021)
32. Klinedinst, N.: Coordinated ifs and theories of conditionals. Synthese 203 (70 2024). https://doi.org/10.1007/s11229-023-04458-y
33. Klinedinst, N., Rothschild, D.: Connectives without truth tables. Nat. Lang. Seman. **20**, 137–175 (2012). https://doi.org/10.1007/s11050-011-9079-5

34. Klinedinst, N.W.: Plurality and possibility. Ph.D. thesis. University of California, Los Angeles (2007). http://citeseerx.ist.psu.edu/viewdoc/download?doi=10.1.1.184.248&rep=rep1&type=pdf
35. Kratzer, A.: What 'must' and 'can' must and can mean. Linguist. Philos. **1**(3), 337–355 (1977). https://doi.org/10.1007/BF00353453
36. Kratzer, A.: The notional category of modality. Words, Worlds, and Contexts: New Approaches in Word Semantics. Berlin: W. de Gruyter (1981). Ed. by Hans-Jürgen Eikmeyer and Hannes Rieser, pp. 39–74. https://doi.org/10.1515/9783110842524-004.
37. Kratzer, A.: Conditionals. Chicago Linguistics Society **22**(2), 1–15 (1986)
38. Daniel Lassiter. Complex sentential operators refute unrestricted Simplification of Disjunctive Antecedents. Semanitcs and Pragmatics 11 (2018). https://doi.org/10.3765/sp.11.9.
39. Lewis, D.: Counterfactuals. Wiley-Blackwell (1973)
40. Loewer, B.: Counterfactuals with disjunctive antecedents. J. Philos. **73**(16), 531–537 (1976)
41. Mandelkern, M.: Talking about worlds. Philos. Perspect. **32**(1), 298–325 (2018). https://doi.org/10.1111/phpe.12112
42. Mandelkern, M.: Bounded modality. Philos. Rev. **128**(1), 1–61 (2019)
43. Mandelkern, M.: Bounded meaning: The dynamics of interpretation. Oxford University Press (2024)
44. Mandelkern, M., Zehr, J., Romoli, J., Schwarz, F.: We've discovered that projection across conjunction is asymmetric (and it is!). Linguist. Philos. **43**(5), 473–514 (2019). https://doi.org/10.1007/s10988-019-09276-5
45. McKay, T., van Inwagen, P.: Counterfactuals with disjunctive antecedents. Philos. Stud. **31**(5), 353–356 (1977). https://doi.org/10.1007/BF01873862
46. Meyer, M.-C.: Generalized free choice and missing alternatives. J. Semant. **33**(4), 703–754 (2016). https://doi.org/10.1093/jos/ffv010
47. Nute, D.: Counterfactuals and the Similarity of Words. J. Philos. **72**(21), 773–778 (1975). https://doi.org/10.2307/2025340
48. Nute, D.: Topics in conditional logic. Springer (1980)
49. Quine, W.V.O.: Methods of Logic. Holt, Rijnehart and Winston, New York (1950)
50. Roberts, C.: Modal subordination, anaphora, and distributivity. PhD thesis (1987)
51. van Rooij, R.: Free choice counterfactual donkeys. J. Semant. **23**(4), 383–402 (2006)
52. Rooth, M., Partee, B.H.: Conjunction, type ambiguity and wide scope or. Proceedings of the first West Coast Conference on Formal Linguistics. Ed. by N. Wiegand D. Flickinger M. Macken, vol. 1. Stanford Linguistics Association Stanford (1982)
53. Santorio, P.: Alternatives and truthmakers in conditional semantics. J. Philos. **115**(10), 513–549 (2018). https://doi.org/10.5840/jphil20181151030
54. Schlenker, P.: Be articulate: a pragmatic theory of presupposition projection. Theor. Linguist. (2008). https://doi.org/10.1515/THLI.2008.013
55. Schlenker, P.: Local Contexts. Semant. Pragmatics **3**, 1–78 (2009)
56. Schwager, M.: Interpreting imperatives. PhD thesis. Frankfurt am Main: Johann Wolfgang Goethe-Universtät dissertation (2006)
57. Stalnaker, R.: A theory of conditionals. In: Harper, W.L. (ed.) Robert stalnaker, and glenn pearce. Springer, pp. 41–55 (1968). https://doi.org/10.1007/978-94-009-9117-0_2.
58. Starr, W.B.: A uniform theory of conditionals. J. Philos. Log. **43**(6), 1019–1064 (2013). https://doi.org/10.1007/s10992-013-9300-8
59. Starr, W.: Conjoining imperatives and declaratives. Proc. Sinn und Bedeutung. **21**(2), 1159–1176 (2018)

60. Veltman, F.: Logics for conditionals. PhD thesis. University of Amsterdam (1985). https://eprints.illc.uva.nl/id/eprint/1834/3/HDS-02-Frank-Veltman.text.pdf
61. Veltman, F.: Defaults in update semantics. J. Philos. Log. **25**(3), 221–261 (1996). https://doi.org/10.1007/BF00248150
62. Warmbrōd, K.: Counterfactuals and substitution of equivalent antecedents. J. Philos. Logic **10**(2), 267–289 (1981). https://doi.org/10.1007/BF00248853.
63. Webber, B., et al.: Anaphora and discourse structure. Comput. Linguist. **29**(4), 545–587 (2003)
64. Willer, M.: Simplifying with Free Choice. Topoi **37**(3), 379–392 (2017). https://doi.org/10.1007/s11245-016-9437-5
65. Woods, M.: Conditionals. Clarendon Press (1997)
66. Thomas Ede Zimmermann: Free choice disjunction and epistemic possibility. Nat. Lang. Seman. **8**(4), 255–290 (2000). https://doi.org/10.1023/A:1011255819284

Questions and Connectives

Tue Trinh[1(✉)] and Itai Bassi[2]

[1] University of Nova Gorica, Nova Gorica, Slovenia
tue.trinh@ung.si
[2] Leibniz-Zentrum Allgemeine Sprachwissenschaft, Berlin, Germany
bassi@leibniz-zas.de

Abstract. We propose a conservative analysis for conditional questions, i.e. those of the form *if* ϕ, Q where ϕ expresses a proposition and Q a question. Our analysis retains the standard interpretation of *if* as a propositional operator and predicts the right intuitions regarding the answers to these questions. Furthermore, we explain why *and* and *or* cannot embed questions the same way *if* does. We show how our account overcomes difficulties faced by previous theories, and discuss some open problems for future research.

Keywords: Questions · Connectives · Co-ordinate Structure Constraint · ATB-movement

1 Introduction

Our starting point is an observation which, to the best of our knowledge, is novel. The observation concerns a difference between *and* and *or* on the one hand, and *if* on the other. The basic intuition we have about these words is that they connect propositions. More technically, these "connectives" denote functions which map two propositions into one: suppose ϕ and ψ express propositions, then ϕ *and* ψ, ϕ *or* ψ, and *if* ϕ, ψ express propositions as well. For present purposes, we will assume the familiar meanings of *and* and *or* as the conjunction and the disjuction, respectively. For *if* we assume the Stalnaker-Lewis-Kratzer 'closest world' analysis (Stalnaker, 1968; Lewis, 1973; Kratzer, 1986).[1]

(1) a. $[\![\text{and}]\!](p)(q) = p \cap q = \lambda w.\ p(w) + q(w) > 1$
b. $[\![\text{or}]\!](p)(q) = p \cup q = \lambda w.\ p(w) + q(w) > 0$
c. $[\![\text{if}]\!](p)(q) = p \mathrel{\Box\!\!\rightarrow} q = \lambda w.$ the p-world closest to w is a q-world

[1] Assuming, of course, that $p(w)$ is either 1 or 0, and the addition sign + has its ordinary meaning. The "p-world closest to w" is the world which differs from w only so much as to make p true. Thus, (*if* ϕ, ψ) says that ψ is true if the actual world were to change minimally as to make ϕ true. Note the singular definite: we make the so-called "limit assumption" that there is exactly one p-world closest to w, for any p and w. For more details see the cited works.

J. Yan et al. (Eds.): TLLM 2024, LNCS 15410, pp. 93–105, 2025.
https://doi.org/10.1007/978-3-031-86054-6_5

The central puzzle which this paper sets out to resolve is this: *if* can connect a proposition with a question, while *and* and *or* cannot.

(2) a. If it's raining, will John come?
b. #It's raining and will John come?
c. #It's raining or will John come?

There is a clear contrast between (2-a) on the one hand and (2-b) and (2-c) on the other.[2] It should be noted, right away, that there is a reading of (2-b) and (2-c) under which these sentences become acceptable. This is the "speech act" reading. Apparently *and* can be used for addition and *or* for revision of speech acts. In written texts, these uses can be brought out more transparently by a non-standard, "creative" punctuation.

(3) a. It's raining. And: Will John come?
'I'm telling you that it's raining. In addition, I'm asking you whether John will come.'
b. It's raining. Or: Will John come?
'I'm telling you that it's raining. On second thought, let me ask you whether John will come.'

Note that the speech act reading is not possible with *if*. An *if*-clause cannot be independent, and thus cannot constitute a speech act. This is reflected in the fact that such creative punctuation as exemplified by (3) cannot be transferred to conditionals.

(4) #If it's raining. Will John come?

The contrast between (4) on the one hand and (3-a) and (3-b) on the other can be derived from the following preference principle regarding the use of the period in English.

(5) The period should end an expression which constitutes a speech act.

Since *it's raining* can constitute a speech act, it can be ended by a period, hence (3-a) and (3-b) are acceptable. In contrast, *if it's raining* cannot constitute a speech act, hence cannot be ended by a period, which is why (4) is odd. Now, is there a way to exclude the speech act reading for *and* and *or*? We think there is. Consider the texts in (6).

(6) a. (i) It's both raining and John will come.
(ii) #It's both raining. And: John will come.
b. (i) It's either raining or John will come.
(ii) #It's either raining. Or: John will come.

[2] As we said in the beginning paragraph, we think this observation is new. It has been noted that a question can be the consequent of a conditional (cf. Isaacs and Rawlins, 2008; Krifka, 2019; Bledin and Rawlins, 2019), but the contrast between *if* and the other two connectives has not been pointed out, as far as we know.

As we can see, the "creative" punctuation does not work when *both* and *either* accompany *and* and *or*, respectively. Given (5), this is evidence that neither *it's both raining* nor *it's either raining* can constitute a speech act. This means that the speech act reading is not available for co-ordinations of the form *both* ϕ *and* ψ and *either* ϕ *or* ψ. Now consider the sentences in (7).

(7) a. #It's both raining and will John come?
b. #It's either raining or will John come?

Speakers we consulted find (7-a) and (7-b) to be substantially worse than (2-b) and (2-c). We believe the reason is that the latter can be rescued by the speech act reading while the former cannot, due to the presence of *both* and *either*.

Let us state the generalization we want to derive.

(8) *If* can connect a proposition and a question, while *and* and *or* cannot

We will propose an analysis which derives (8) and, at the same time, maintains that *if* is propositional in both arguments. Our discussion will focus on polar questions, but the proposal generalizes to constituent questions, as will be seen.

2 Two Previous Accounts

Henceforth we will use the term "conditional questions", short CQs, to refer to sentences such as (2-a). We will now discuss two (kinds of) analyses of CQs that have been proposed in the literature.

2.1 The Tripartition Analysis

According to the "tripartition analysis", proposed by Groenendijk and Stokhof (1997), CQs partition the context set into three cells.[3] For example, the CQ in (2-a), reproduced in (9), would result in the context set in (10).

(9) If it's raining, will John come?

(10) Context set induced by (9)

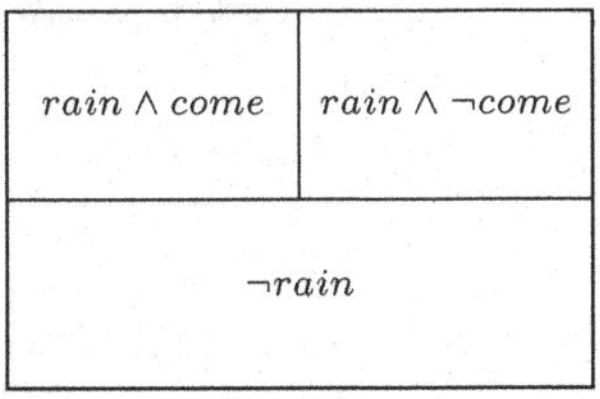

[3] The "context set" is the set of possible worlds which represents the conjunction of all mutual assumptions of all discourse participants (cf. Stalnaker, 1978).

The question thus presents the addressee with three choices: (i) affirm the antecedent and saying "yes" to the consequent (11-a); (ii) affirm the antecedent and saying "no" to the consequent (11-b); and (iii) denying the antecedent (11-c).

(11) a. It's going to rain and John will come
b. It's going to rain but John will not come
c. It's not going to rain

We agree that the sentences in (11) can be used as responses to (9). To the extent that this is true, the tripartition analysis does have merits. However, we believe that there is a clear sense in which the responses in (11) do not directly address the question. They seem to target not the question but, rather, its presupposition, triggered by *if*, that rain is possible but not certain (von Fintel, 1999). The question asks whether John will come in the closest rain-world, i.e. whether John will come if the actual world were to change minimally so that it is raining. It does not ask whether it's going to rain, or whether John will come. However, these are the questions which are answered by the sentences in (11). Our claim, therefore, is that the tripartition analysis overgenerates: it includes sentences that should not be included.

The tripartition analysis, we believe, also undergenerates. Our intuition, which is shared by native speakers we have consulted, is that both *yes* and *no* are perfect answers to (9), and are interpreted as indicated in (12-a) and (12-b).

(12) If it's raining, will John come?
a. Yes (= 'if it's raining, John will come')
b. No (= 'if it's raining, John will not come')

But according to the tripartition analysis, (12-a) and (12-b) are not answers to (9), reproduced in (12), at all. In fact, this analysis implies that CQs are not yes/no questions. A yes/no question has two answers, which means it partitions the context set into two cells, not three. Thus, the tripartition analysis undergenerates: it excludes sentences that should not be excluded.

Last but not least, the tripartition analysis has nothing to say about the contrast between *if* on the one hand and *and* and *or* on the other (Velissaratou, 2000). Why should (7-a) not be a well-formed question which partitions the context set into a $rain \wedge come$ cell and a $rain \wedge \neg come$ cell, for example?

We take the above considerations to be sufficient grounds to look for another analysis.

2.2 The Context Update Analysis

The "context update" analysis, proposed in various forms by Isaacs and Rawlins (2008); Krifka (2019); Bledin and Rawlins (2019), take the interpretation of CQs to be a two-step process. First, the initial context c is updated to a "temporary" context c' by the *if*-clause. Then the question in the consequent is asked with respect to this local context c'.

(13) c + If [$_p$ it's raining], [$_Q$ will John come?]
(i) update c with p
$\rightarrow$ resulting in c'
$= \lambda w.\ w \in c \wedge$ it's raining in w
(ii) update the output of (i) with Q
$\rightarrow$ resulting in a partition of c'
$= \lambda w.\ \lambda w'.\ w, w' \in c' \wedge$ (it's raining in $w \leftrightarrow$ it's raining in w')

This analysis seems more promising than the tripartition analysis. A CQ does seem to work in just the way described. In uttering the CQ in (2-a), we are not asking whether John will come in the actual context. Instead, we are asking whether he will come in the hypothetical scenario in which it is raining. Also, we can see how *yes* and *no* can be interpreted under this analysis: *yes* means John will come in the hypothetical scenario, and *no* means he won't come in the hypothetical scenario. Thus, the context update analysis is also superior to the tripartition analysis in that it predicts *yes* and *no* to be answers to CQs and, furthermore, to have the intuitively correct interpretation.

The context update analysis also assimilates CQs and regular conditionals with respect to the phenomenon of "denying the antecedent".

(14) A: If it's raining, John will not come.
B: It's not going to rain.

(15) A: If it's raining, will John come?
B: It's not going to rain.

Intuitively, B is doing the same thing in both (14) and (15), which is claiming that the update to be performed as instructed by the *if*-clause is not realistic. What B says is that all worlds in the context set are non-rain worlds. Thus, updating this context with *it's raining* will result in the contradiction, and will not be a pragmatically felicitous move (Stalnaker, 1978).

Another virtue of the context update analysis is that it squares with facts about presupposition projection. Consider the assertion in (16-a) and the question in (16-b).

(16) a. The king of France is bald. $\rightsquigarrow$ France has a king
b. Is the king of France bald? $\rightsquigarrow$ France has a king

Both of these sentences presuppose that France has a king. Now, we know that the presupposition of *if* ϕ, ψ_p, where p is the presupposition of ψ, is $[\![\phi]\!] \subseteq p$ (Karttunen, 1973; Heim, 1990). In other words, the *if*-clause "filters out" the presupposition of the consequent. This means that if $[\![\phi]\!] = p$, then *if* ϕ, ψ_p will have the trivial presupposition that $p \subseteq p$, which is to say that it will not presuppose anything. That this is the case is evidenced by (17).

(17) If France has a king, the king of France is bald $\rightsquigarrow \top$

The update semantics proposed for conditionals predicts exactly this projection behavior. Now consider (18).

(18) If France has a king, is the king of France bald? $\rightsquigarrow \top$

Intuitively, (18) presupposes nothing, in exactly the same way that (17) presupposes nothing. This means the same update semantics holds for *if*-clauses, both in regular conditionals and in conditional questions. A good result.

So why should we not adopt the context update analysis for CQs? Well, it does not answer the question we want to answer, namely why *if* is different from *and* and *or*. According to the standard update semantics for *and* and *or*, the presupposition projection behavior of ϕ *and* ψ is similar to that of *if* ϕ, ψ and the presupposition projection behavior of ϕ *or* ψ is similar to that of *if* $\neg\phi$, ψ. Thus, the local context for the second conjunct is the first conjunct, and the local context for the second disjunct is the negation of the first disjunct. This is evidenced by the fact that neither (19-a) and (19-b) presupposes anything.

(19) a. France is both a monarchy and the king of France is bald
b. France is either not a monarchy or the king of France is bald

But then it's not clear why *and* and *or* cannot embed questions. Why should (20-a) and (20-b) not be well-formed and presuppose nothing in the same way as (19-a) and (19-b), respectively?

(20) a. #France is both a monarchy and is the king of France bald?
b. #France is (either) not a monarchy or is the king of France bald?

Let us consider another analysis.

3 Proposal

This section presents our explanation of the generalization in (8). We show that it reduces to a well-known fact, namely the Coordinate Structure Constraint, assuming that all connectives are propositional.

3.1 Analysis of Polar Questions

We implement the semantics of (matrix) polar questions along the line of several well-known analyses (cf. Bennett, 1977; Higginbotham, 1993; Krifka, 2001a; Guerzoni, 2004). Specifically, we assume that polar questions contain a covert *whether* which moves to [Spec, C] from the edge of a propositional constituent, leaving a trace and creating a λ-abstract. We take the head-movement of *will* which results in subject-auxiliary-inversion to be semantically inconsequential, and will not represent it in logical form.

(21) a. will John come?

b.

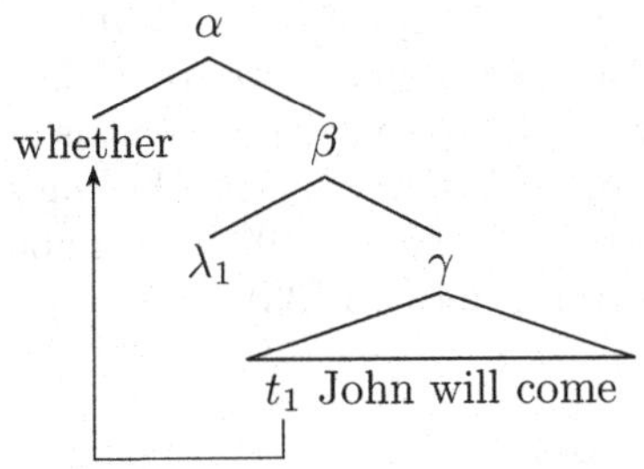

We define two functions from propositions to propositions: YES $=_{\text{def}} \lambda p.\ p$, and NO $=_{\text{def}} \lambda p.\ \neg p$. Let us call YES the "positive polarity" and NO the "negative polarity", and write "$pol(f)$" to mean f is a polarity, i.e. f is either YES or NO. We want the interpretation of (21-a) to be the set in (22).

(22) {YES(John will come), NO(John will come)}
= {John will come, John will not come}

This means α in (21-b) should be of type $\langle st, t\rangle$. The semantic types of the constituents are presented in (23). Note that the trace of *whether* is of the "polarity" type, i.e. $\langle st, st\rangle$.

(23)

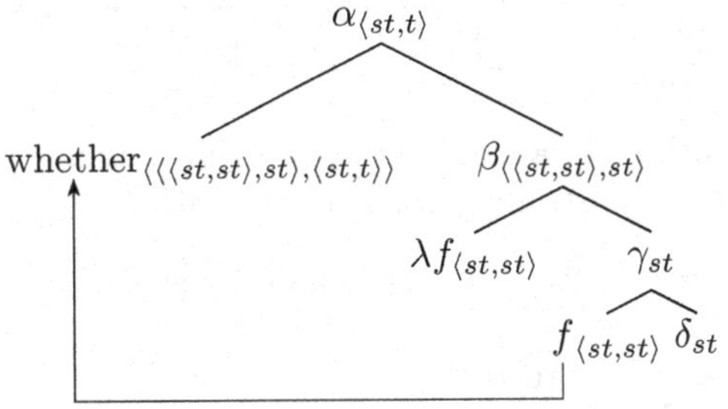

Here are the meanings of the constituents. We invite the readers to verify that applying this analysis to (21-a) will yield the set in (22), as desired.

(24) a. $[\![\text{whether}]\!] = \lambda Q_{\langle\langle st,st\rangle,st\rangle}.\ \lambda p_{st}.\ \exists f_{\langle st,st\rangle}.\ pol(f) \wedge p = Q(f)$
b. $[\![\beta]\!] = \lambda f.\ f([\![\delta]\!])$
c. $[\![\alpha]\!] = \lambda p.\ \exists f.\ pol(f) \wedge p = [\![\beta]\!](f)$
$= \{[\![\delta]\!], \neg[\![\delta]\!]\}$

3.2 Deriving the Observations

Types and Locality – Let us now turn to the main topic of this paper: conditional questions (CQs) and the difference between *if* and the other two connectives. We start with the claim in (25).

(25) All connectives are propositional

In other words, both arguments of *if*, as well as those of *and* and *or*, are of type $\langle s, t\rangle$, and the interpretation of the three connectives are as presented in (1). We propose that CQs are just polar questions whose prejacent is a conditional. In particular, there is a covert *whether* which moves from the edge of the main clause, i.e. the consequent, leaving a trace (Bennett, 1977; Higginbotham, 1993; Krifka, 2001a; Guerzoni, 2004). The logical form of (33) would be (26-b), whose interpretation would the set in (26-c).

(26) a. if it's raining, will John come
b. $[_\alpha$ whether λ_1 [[if it's raining] [t_1 John will come]]]]
c. $[\![\alpha]\!] = \{$rain $\Box\!\!\rightarrow$ John comes, rain $\Box\!\!\rightarrow$ $\neg$John comes$\}$
= {if it's raining John will come, if it's raining John won't come}

Note that movement of *whether* violates no locality constraints because the *if*-clause is a subordinate clause, which is an adjunct. As evidenced by (27) below, a wh-phrase can move across an *if*-clause, just like it can move across a non-sentential adjunct.

(27) a. that's the movie which [[if it's raining] I would watch *t*]
b. that's the movie which [[surely] I would watch *t*]

But note that wh-movement out of a conjunct or a disjunct is not possible, due to the Coordinate Structure Constraint (CSS) (Ross, 1967).

(28) a. *that's the movie which [[it was raining] and [I watched *t*]]
b. *that's the movie which [[I read the book] or [I watched *t*]]

We are now in the position to resolve the central puzzle of the paper. First, let us address the fact that *and* and *or* cannot connect a proposition and a question. Consider, again, the (unacceptable) sentence in (2-b), reproduced here in (29).

(29) #It's raining and will John come?

What are the possible analyses of (29)? Suppose *whether* moves within the second conjunct, as depicted in (30).

(30) $[_{st}$ it's raining] and $[_{st,t}$ whether *t* John will come]

Given (25), (30) is a case of type mismatch: the left argument of *and* is a proposition (type *st*), but the right argument is a question, i.e. a set of propositions (type $\langle st, t\rangle$). Now, suppose *whether* moves to scope above both conjuncts, as depicted in (31).

(31) [whether $[_{st}$ it's raining] and $[_{st}$ t John will come]]

In this case, both arguments of *and* are of the right type. However, the movement of *whether* violates the CSC. We do not see any other landing site for *whether*. Since both possible parses of (29) are problematic, the sentence is unacceptable.

A completely parallel story can be told for *or*. The possible parses for the unacceptable (2-c), reproduced in (32), are given in (32-a), which is a type mismatch, and (32-b), which violates the CSC.

(32) It's raining or will John come?
  a. $[_{st}$ it's raining] or $[_{st,t}$ whether t John will come] → *type
  b. [whether $[_{st}$ it's raining] or $[_{st}$ will t John t_{will} come]] → *CSC

Answers – Our account of CQs makes the correct predictions regarding the answers to these questions. As seen from (26), we predict (33-a) and (33-b) to be the two answers to (33).

(33) If it's raining, will John come?
  a. If it's raining, John will come = 'yes'
  b. If it's raining, John will not come = 'no'

If we make the completely natural assumption that *yes* and *no* associate with assigning the values YES and NO to the polarity function, respectively, we get the result that the *yes* answer to (33) has the meaning of (33-a), and the *no* answer the meaning of (33-b). This result, we believe, accords with intuition: (33-a) and (33-b) are felt to be the two answers to (33). Furthermore, we predict the sentences in (11), reproduced below in (34-a) to (34-c), not to be congruent answers to (34).

(34) If it's raining, will John come?
  a. It's going to rain and John will come
  b. It's going to rain but John will not come
  c. It's not going to rain

This, we believe, is also intuitively correct. All of these responses have a feel of "presupposition denial". They seem to say that there is something infelicitous about the question. And this is exactly what we predict. Look again at the two answers (33-a) and (33-b). Both of them are conditionals of the form *if* ϕ, ψ. This means both of them presuppose that rain is possible but not certain: some worlds in the context set are rain worlds, and some of them are non-rain worlds (von Fintel, 1999). This is thus the presupposition of the question. The assertions in (34-a) to (34-c) all deny this presupposition: (34-a) and (34-b) entails that all worlds in the context set are rain worlds, and (34-c) entails that all worlds in the context set are non-rain worlds.

4 Loose Ends

4.1 Factivity Effects

Consider the contrast between (35-a) and (35-b).[4]

(35) a. If it's raining, will John come?
b. #Because it's raining, will John come?

Suppose *because*, just like the other connectives, is propositional in the sense that both of its arguments are of type *st*. Then the only parse for (35-b) which does not incur type mismatch is (36).

(36) [$_\alpha$ whether λ_1 [[because it's raining] [t_1 John will come]]]]

Note that *because* comes with a factive presupposition for both of its arguments. This means that movement of *whether* in (36) is one out of a "factive island", which in turns means that it is not available. The contrast below provides independent evidence of this fact.

(37) a. what do you think [$_\alpha$ John saw *t*]

b. #what do you know [$_\alpha$ John saw *t*]

4.2 Constituent Questions

We have discussed polar questions, but the account we propose should generalize to constituent questions. First, note that the contrast between *if* and the other two connectives obtains for constituent questions as well.

(38) a. If it's raining, who will come?
b. #It's (both) raining and who will come?
c. #It's (either) raining or who will come?

We have interpreted *whether* as 'which polarity f is such that ...'. Let us interpret *who* in the same manner, namely as 'which person x is such that...'. The semantics for *who* is given in (39-b), and that for *whether* is reproduced in (39-a) for easy comparison.

(39) a. $[\![\text{whether}]\!] = \lambda Q_{\langle\langle st,st\rangle,st\rangle}.\ \lambda p_{st}.\ \exists f_{\langle st,st\rangle}.\ pol(f) \wedge p = Q(f)$
b. $[\![\text{who}]\!] = \lambda P_{\langle e,st\rangle}.\ \lambda p_{st}.\ \exists x_e.\ person(x) \wedge p = P(x)$

[4] A qualification is in order. We believe that there is a "speech act embedding" reading in which (35-b) is felicitous, namely 'because it is raining, I am going to ask you whether John will come'. Similarly for sentences such as *since it's raining, will John come?*, or *now that we know it's raining, will John come?* (we thank an anonymous reviewer for drawing out attention to these data points.) We have nothing to say about this reading within the confines of this paper. For analyses of embedded speech acts see Krifka (2001b, 2014).

The logical form of (38-a) is (40-a), and its denotation is the set in (40-b). The congruent answers are predicted to be such sentences as those in (40-c).

(40) a. who λx [if it's raining [t_x will come]]

b. {rain $\square\!\!\rightarrow$ John comes, rain $\square\!\!\rightarrow$ Mary comes, rain $\square\!\!\rightarrow$ Sue comes,...}

c. (i) If it's raining, John will come
(ii) If it's raining, Mary will come
(iii) If it's raining, Sue will come

We believe this is the intuitively correct result. As for the unacceptable (38-b), the logical form would be (41-a), the denotation would be the set in (41-b), and the congruent answers sentences such as those in (41-c).

(41) a. who λx [[it's raining] and [t_x will come]]

b. {rain $\wedge$ John comes, rain $\wedge$ Mary comes, rain $\wedge$ Sue comes...}

c. (i) It's raining and John will come
(ii) It's raining and Mary will come
(iii) It's raining and Sue will come

As far as we can see, there is nothing wrong with (41-b) and (41-c). The unacceptability of (38-b) is thus due to a syntactic constraint, i.e. the CSC. And what we just said about (38-b) can obviously be extended to (38-c), as the above arguments can be reproduced with *or* replacing *and*.

4.3 Ordering

The above discussion does not depend on the ordering of the arguments of the connectives: an *if*-clause is an adjunct whether it precedes or follow the main clause, and movement out of a co-ordinated clause is excluded whether it is the right or the left argument of the relevant connective. We thus predict that ordering of the clauses should not matter. This prediction is borne out, as evidenced by the contrasts in (42) and (43).

(42) a. Will John come if it's raining?
b. #Will John come and it's raining?
c. #Will John come or it's raining?

(43) a. Who will come if it's raining?
b. #Who will come and it's raining?
c. #Who will come or it's raining?

4.4 ATB-Movement and Open Problems

It is a well-known fact that questions can be conjoined, as illustrated in (44). For this case, we make the correct prediction, as nothing prevents *whether* to ATB-move out of both co-ordinate clauses, as shown in (44-a). The set of answers is predicted to be that in (44-b).

(44) is it raining and will John come?

a. whether λ_1 [[t_1 it is raining] and [$_\alpha$ t_1 John will come]]

b. {rain $\wedge$ John comes, $\neg$rain $\wedge$ $\neg$John comes}

The readings of *yes* and *no* seem to support the analysis. Specifically, the *yes* answer seems to confirm both conjuncts, while the most natural reading of *no* seems to be that which denies both conjuncts.

(45) Is it raining and will John come?

a. Yes. $\rightsquigarrow$ it's raining and John will come
b. No. $\rightsquigarrow$ it's not raining and John won't come

However, we admit that the *no* answer also has the 'not both' reading, i.e. $\neg$(rain $\wedge$ John comes). We do not yet see how this reading of *no* can be explained given the logical form in (44-a). This is thus an open problem for us.

Another open problem concerns disjunction. As it stands, the analysis does not rule out ATB-movement from both disjuncts, which means we predict (46) to be possible with (46-a) as its logical form and (46-b) its denotation.

(46) #Is it raining or will John come?

a. whether [[t_1 it is raining] or [$_\alpha$ t_1 John will come]]

b. {rain $\vee$ John comes, $\neg$rain $\vee$ $\neg$John comes}

As indicated by the # sign, we believe (46) is not acceptable under the intended reading. There is, of course, the "alternative" reading of (46) under which this question denotes the set of proposition in (47), but this reading is not one that is expressed by the logical form in (46-a).[5]

(47) {it's raining, John will come}

We leave these issues to future research.

5 Conclusion

It has been observed that *if* can embed a proposition as antecedent and a question as consequent. We pointed out the fact that this is not possible with *and* and *or*, which has not been highlighted in the literature. We then provide an explanation for this fact. Our account is based on independently motivated claims about syntax and semantics: (i) connectives are propositional; (ii) *if* is a subordinator while *and* and *or* are co-ordinator; (iii) matrix polar questions contain a covert *whether* which quantifies over polarities and undergoes wh-movement; (iv) wh-movement is subject to the Co-ordinate Structure Constraint. We show that our

[5] We will remain agnostic about the analysis of alternative questions in this note.

analysis does justice to intuitions about the answers to conditional questions in contrast to previous theories. Finally, we discuss some open problems regarding ATB-movement of wh-phrases which we leave to future research.

Acknowledgement. We thank Anton Benz, Luka Crnic, Dan Goodhue, Roland Hinterhölzl, and Manfred Krifka for valuable input and discussion. The first author, Tue Trinh, is financially supported by the Slovenian Research Agency (ARIS) project no. J6-4615. All errors are our own.

References

Bennett, M.: A response to Karttunen. Linguist. Philos. **1**, 279–300 (1977)

Bledin, J., Rawlins, K.: What ifs. Semant. Pragmat. **12**, 1–62 (2019)

von Fintel, K.: NPI licensing, Strawson entailment, and context dependency. J. Semant. **16**, 97–148 (1999)

Groenendijk, J., Stokhof, M.: Questions. In: Van Benthem, J., Ter Meulen, A. (eds.) Handbook of Logic and Language, pp. 1055–1124. Elsevier, Amsterdam (1997)

Guerzoni, E.: Even-NPIs in yes/no questions. Nat. Lang. Seman. **12**, 319–343 (2004)

Heim, I.: Presupposition projection. In: Presupposition, Lexical Meaning and Discourse Processes: Workshop Reader, ed. Rob van der Sandt. University of Nijmegen (1990)

Higginbotham, J.: Interrogatives. In: The View from Building 20, ed. Kenneth Hale and Samuel Jay Keyser, 195–228. MIT Press, Cambridge (1993)

Isaacs, J., Rawlins, K.: Conditional questions. J. Semant. **25**, 269–319 (2008)

Karttunen, L.: Presupposition of compound sentences. Linguist. Inq. **4**, 169–193 (1973)

Kratzer, A.: Conditionals. In: von Stechow, A., Wunderlich, D. (eds.) Semantics: An International Handbook of Contemporary Research, pp. 651–656. Mouton de Gruyter, Berlin (1986)

Krifka, M.: For a structured account of questions and answers. In: Audiatur Vox Sapientiae. A Festschrift for Arnim von Stechow, ed. Caroline Fery and Wolfgang Sternefeld, pp. 287–319. Akademie Verlag, Berlin (2001a)

Krifka, M.: Quantifying into question acts. Nat. Lang. Semant. **9**, 1–40 (2001b)

Krifka, M.: Embedding illocutionary acts. In: Roeper, T., Speas, M. (eds.) Recursion: Complexity in Cognition. STP, vol. 43, pp. 59–87. Springer, Cham (2014). https://doi.org/10.1007/978-3-319-05086-7_4

Krifka, M.: Indicative and subjunctive conditionals in commitment spaces. In: Proceedings of the 22nd Amsterdam Colloquium, pp. 248–258 (2019)

Lewis, D.: Counterfactuals. Basil Blackwell, Oxford (1973)

Ross, J.: Constraints on Variables in Syntax. Doctoral Dissertation, Massachusetts Institute of Technology, Cambridge (1967)

Stalnaker, R.: A theory of conditionals. In: Studies in Logical Theory, ed. Nicholas Rescher, pp. 315–332. Blackwell (1968)

Stalnaker, R.: Assertion. Syntax Semant. **9**, 315–332 (1978)

Velissaratou, S.: Conditional questions and which-interrogatives. Master's thesis, University of Amsterdam (2000)

Variation in Conditional Perfection: A Comparative Study of English/German Versus Mandarin Chinese

Alexander Wimmer[1(✉)] and Mingya Liu[2]

[1] Institute of Linguistics, Universität Stuttgart, Keplerstraße 17, 70174 Stuttgart, Germany
alexanderwimmer2012@gmail.com

[2] Department of English and American Studies, Humboldt-Universität zu Berlin, Unter den Linden 6, 10099 Berlin, Germany
mingya.liu@hu-berlin.de

Abstract. Conditional perfection (CoP) refers to the phenomenon where a bare conditional sentence (*if p, q*) receives a bi-conditional (*iff p, q*) interpretation. This paper pursues two main goals: (i) to probe into possible grammatical factors (dis)favoring a CoP-interpretation of a given conditional *if p, q*; (ii) to shed some light on the crosslinguistic picture, checking whether the same grammatical manipulations are equally (ir)relevant for a conditional's CoP-interpretation across two typologically unrelated groups of languages: Chinese on the one hand and English/German on the other. We consider three manipulations that are equally CoP-(dis)favoring in both groups, and two manipulations that seem to favor CoP only in English/German, but not in Chinese. The main finding is that some, but not all grammatical manipulations share the same pragmatic effect in the two groups under consideration.

Keywords: Conditional perfection · Implicatures · Crosslinguistic comparison

1 Introduction

Bare conditionals of the form *if p, q* often convey the truth of the antecedent p to be not only *sufficient*, but also *necessary* for the consequent q to be true:[1]

(1) If you mow the lawn, I'll give you $5. [17]
⇝ If **and only if** [$_p$ you mow the lawn], [$_q$ I'll give you $5].

[17] follow Lauri Karttunen in calling this interpretive tendency ***conditional perfection*** **(CoP)**, an "invited inference" excluding the possibility that 'you' do something other than mowing the lawn and still get $5.

[1] There are degrees of bareness for bare conditionals. Chinese bare conditionals may lack an *if*-like complementizer, and even the 'then'-particle *jiu* may be absent [26,40,43]. Taking this into account, a more accurate way of representing bare conditionals would be (*if*) *p*, (*then*) *q*. We thank Xuetong Yuan (pc) for discussion on this point.

J. Yan et al. (Eds.): TLLM 2024, LNCS 15410, pp. 106–126, 2025.
https://doi.org/10.1007/978-3-031-86054-6_6

There are different ways of describing CoP. The paraphrase in (1) adds an *only if* to the *if*. This is not the only paraphrase available. CoP has been described in at least the following three ways; '$\rightsquigarrow$' stands for 'implies' (in a very general sense).

(2) *if p, q* $\rightsquigarrow$ if p, q &
  a. only if p, q
  b. if not p, not q 'denying the antecedent'
  c. if q, p 'affirming the consequent'

We will not sharply distinguish between these three forms, even though the difference can matter, as shown by [56].[2]

CoP is widely seen as a *pragmatic* (strengthening) inference [4,14,17,23,29], with some recent work focusing on theoretical and experimental modelling [9,12,21]. However, there has to our knowledge been little research on the crosslinguistic picture, aside from a few exceptions including [62] on Russian and [63] on Mandarin Chinese. What also still seems unclear is the full range of factors that (dis)favor CoP-readings, or don't affect them at all. In the following, we will refer to such factors as **CoP-(dis)favoring** or **CoP-neutral** factors.

This paper is intended as another step towards filling the gaps just described. In line with this aim, our broader research questions are

(i) what CoP-(dis)favoring factors there are
(ii) how they differ across languages

In approaching these questions, we offer a preliminary case study comparing CoP-inferences in Chinese vs. English and German regarding a selection of potentially relevant factors. It is clear that what we offer is far from comprehensive in typological terms. Still, we find great appeal in comparing two groups of languages this far apart: while differences are to be expected, showcasing *language particulars*, similarities can be taken as potential candidates for what might end up being *language universals*.

Our comparisons are drawn against the backdrop of the following three working hypotheses:

(H1) **Variability**:
Even within a given language, CoP varies considerably due to *grammatical* (construction-specific) and *pragmatic* (contextual) properties

(H2) **Gradability**:
CoP is gradable (or 'cumulative'): the more CoP-favoring factors come together, the stronger the CoP-inference becomes

[2] From a psycholinguistic viewpoint, (2-b) and (2-c) are problematic paraphrases for testing whether participants in an experiment 'perfect' a given conditional or not [42]: (2-b) adds negation, which may cause processing difficulties; (2-c) comes with the undesirable side effect of reversing the temporal relation between p and q, an observation dating back to [44].

(H3) **Link to the QUD**, following [14]:
CoP-favoring factors are linked to a specific **Question under Discussion (QUD)** [48]

(H2) predicts the following correlations: for a given conditional C, the more favoring factors come together, the stronger a CoP-inference becomes for C, and the more *dis*favoring factors come together, the *weaker* that inference becomes. This strength may be empirically testable in terms of the (in)felicity of a CoP-denying continuation following the utterance of C. This may seem overly strong, but serves as a working assumption; the combination of more than one candidate factor will not be our main concern here anyway.[3]

(H3) assumes [14]'s QUD-approach to CoP, which links the presence of CoP to a certain type of QUD, which we refer to as QUD1, and its absence to another type, which we refer to as QUD2:

(3) a. QUD1 (CoP-favoring): *under which conditions q?*
b. QUD2 (CoP-neutral): *what if p?*

QUD1 asks for the conditions that verify the consequent q. If a conditional *if p, q* is interpreted as a complete (or *exhaustive*) answer to that question [19], we get CoP: if there were q-verifying conditions other than p, they would have been mentioned. QUD2, by contrast, asks for the consequences that follow from p being true. This question has nothing to do with alternative conditions making q true, so we characterize it as 'CoP-neutral' in (3-b).

(H3) promotes the QUD to the overarching variable underlying all other CoP-favoring vs. -disfavoring factors there may be, so it 'sees' a clear pragmatic dimension even to the grammatical manipulations we are going to focus on. CoP-favoring factors are taken to be indicative of, and more compatible with, QUD1; the same goes for CoP-disfavoring factors in relation to the CoP-neutral QUD2.[4]

---

[3] There is reason to believe that at least some factors, whether favoring or disfavoring, are stronger than others. A fairly clear case is the step from *if* to *even if*, a manipulation that can be expected to 'wipe out' CoP altogether, however many favoring factors might be present. For example, there is no obvious way in which the following variation of (1) could be thought of as exhibiting CoP, i.e. as conveying the antecedent to be the only condition under which the consequent holds:

(i) **Even** if you mow the lawn, I'll give you \$5.
$\not\leadsto$ If **and only if** you mow the lawn, I'll give you \$5.

[4] Combining (H2) and (H3), one may be tempted to say: the more favoring factors come together, the stronger the link to QUD1, and the stronger the CoP-inference; the reverse pattern is predicted for the disfavoring factors. This correlative hypothesis inherits the problems already faced by (H2) alone: if a conditional is preceded by QUD1, it is unclear whether it can be further perfected by activating other CoP-favoring factors. So again, some factors – be it a CoP-favoring QUD1 or CoP-disfavoring *even if* from footnote (see footnote 3) – seem to have a stronger impact than others.

A 'rule of thumb' to identify the QUD is focus placement. This follows from [49]'s *question-answer constraint*: informally speaking, focus highlights that piece of information which is asked for in a preceding question. The constraint is central to [12]'s experimental investigation of the QUD-approach to CoP: focus on p creates a link to QUD1 in that it relates to a question about p; by contrast, it is *incongruent* with QUD2, which asks about q, not p. It is focus on q that is linked to QUD2, and incongruent with QUD1, for analogous reasons. These associations are sketched in Table 1 below, where we follow the convention of indicating focus-marking with the subscript 'F'.[5]

Table 1. QUD-dependent focus placement in conditionals

	[QUD1 under which conditions q?]	[QUD2 what if p?]
if p_F, q	✓	ϟ
if p, q_F	ϟ	✓

In approaching the research questions stated above, we will proceed as follows: in Sect. 2, we first look at **similarities** between Chinese and E/G (English/German). In doing so, we zoom in on three kinds of grammatical manipulations that have the same effect on CoP in both language groups: a CoP-neutral, a CoP-disfavoring and a CoP-favoring one. (4) illustrates each manipulation in turn:

(4) a. If it rains, stay inside!
b. If it rains, will she stay?
c. Only if it rains will she stay.

The sentence in (4-a), a *conditional[ized] imperative* [10,37,52] or *imperative conditional* [47], has an imperative rather than an indicative clause as its consequent: a CoP-neutral mood switch. (4-b) is a conditionalized polar question: a construction that has been claimed not to come with CoP [11,29], and will hence be classified as CoP-disfavoring. (4-c), finally, is an *only if* conditional: a construction that is intuitively more prone to CoP than a bare *if*-conditional, hence intuitively classifiable as a CoP-favoring construction.

Having looked at interlinguistic similarities, Sect. 3 discusses two CoP-related **differences** between Chinese and E/G. Again, we look at two grammatical manipulations, exemplified in (5); the subscript 'F' in (5-b) indicates (intonational) focus on the particle *then*.

[5] In [13]:81, it is primarily the position of the *if*-clause (left vs. right) whose relation to the QUD is considered. This positional difference, which is clearly not unrelated to focus-marking, is neglected here, but see Sect. 3.1 for further discussion.

(5) a. (i) If it rains, she'll stay.
(ii) She'll stay if it rains.
b. If it rains, $then_F$ she'll stay.

In (5-a) , the antecedent-clause alternately appears to the left and to the right of the consequent clause. [8] claims 'right' antecedents to favor, and 'left' ones to disfavor, (what is now known as) a CoP-reading: a CoP-enforcing strategy that is practically unavailable in Chinese [46]. The second and final difference as exemplified by (5-b) pertains to (prosodic) focus on conditional *then* [50, 60]. [50] treats conditionals with focused *then* as 'perfected'. Again, the grammar of Chinese proves to be more restrictive in this regard, disallowing focus on *then.* In sum, then, we think of right-adjoined antecedents as well as focus on conditional *then* as CoP-favoring factors that are available in E/G, but not in Chinese.

In short, the paper is organized as follows: Sects. 2 and 3 discuss interlinguistic similarities and differences, respectively, and Sect. 4 concludes the paper.

2 Similarities

Certain grammatical manipulations seem to have just the same effect on a conditional's 'perfection' in either of the two groups of languages under consideration here. We zoom in on the three manipulations mentioned above. The first one is the 'mood switch' of the consequent q from indicative to imperative, which leads to the configuration *if p, q*-IMP; the second one is the simple addition of *only* to *if*, and the third one is the formation of a polar question (*if p, q?*). In both language groups, the three manipulations will be classified as CoP-neutral, -favoring and -disfavoring, respectively.

2.1 Imperativized Consequents as CoP-Neutral

We see the conditional consequent's 'imperativization' as a CoP-neutral manipulation. The construction resulting from this manipulation is what we follow [47] in calling an *imperative conditional* (IC). (6) provides two examples, one from English and one from Chinese.[6]

(6) a. If it rains, stay.
b. Yaoshi xiayu, jiu liuxia ba.
if rain JIU stay IMP

Leaving open the possibility that some Mandarin imperatives lack special morphology altogether, the pattern for ICs boils down to the following:

[6] As the reader might have expected, Chinese imperative clauses don't come with dedicated verbal morphology, but the subjects tend to be omitted like in English and German. They may contain the sentence-final particle *ba*, which is tentatively glossed as an imperative marker in (6-b), but has a variety of other uses as well; see e.g. [61].

(7) *if p, q*-IMP

Under the QUD-approach to CoP, an IC's CoP-neutrality means that whether an IC is perfected or not depends on the QUD in the way outlined above:

(8) An *imperative conditional IC* of the form *if p, q*-IMP is perfected under a CoP-favoring QUD1 (*under which conditions q?*), but not under a CoP-neutral QUD2 (*what if p?*).

What does a QUD1 have to look like for an IC? This QUD shares the same consequent with the IC that serves as its answer, so it should have a consequent in the imperative. But this leads to ungrammaticality, as can be seen in (9): (9-b) is a failed attempt at forming an 'imperativized' QUD1 for the IC in (9-a).

(9) a. Stay if it rains.
b. *Under which conditions stay?

There is work on the degradation of imperatives under questions which we cannot do justice to here; cf. e.g. [35,45,54]. At an intuitive level, it is clear that the QUD1 has to involve a modal instead of an imperative, which lends some support to a modal treatment of imperatives such as [35]'s:

(10) Under which conditions {may I, do I have to} stay?

This QUD varies between an existential and a universal modal. This variation touches on the well-known fact that imperatives vary between *weak* and *strong* readings, approximately: permissions and commands; see [15,36] for overviews. In what follows, we are going to leave aside weak readings and existentialized QUDs, whose interaction with CoP-inferences is discussed in [59]. Instead, we focus on the hypothesis stated in (8): ICs are sometimes, but not always perfected, and this variability can be ascribed to the respective QUD. For this purpose, it remains to be established how to detect CoP in an IC. [59] proposes that under a *strong* reading for the imperative consequent q-IMP,[7] CoP has the effect of a conditionalized *permission* not to perform the action described by q-IMP, the condition being that the antecedent does not hold:

(11) A perfection-reading for an IC of the form *if p, q*-IMP is interpreted such that ¬q is **permitted** in all ¬p-cases.

Intuitions about such readings are fairly subtle, and call for experimental investigation. For now, we take both Chinese and English ICs to come with the 'permissive' flavor stated in (11) when they are preceded by a CoP-favoring QUD1:

(12) **QUD1: Under which conditions do I have to stay?**
a. Stay if it rains.

[7] We take this strong reading to be the default; see again [35] a.o.

b. Yaoshi xiayu, jiu liuxia ba.
if rain then stay IMP
⇝ If it doesn't rain, you are permitted not to stay.

Since this permissive inference is by hypothesis a CoP-inference, we predict it to be absent following a CoP-neutral QUD2:[8]

(13) **QUD2: What if it rains?**

a. If it rains, stay.

b. Yaoshi xiayu, jiu liuxia ba.
if rain JIU stay IMP

With [29], one may raise the following objection against the predicted absence of CoP from (13): in a dialogue as in (13), would a speaker reply the way they do if they wanted the addressee to stay *no matter what*? If such an *unconditional* reading is ruled out, we must conclude some form of CoP to be at play even in (13), and it might well be a (pragmatic) *presupposition* instead of an *implicature* in this case. We leave a clarification of these subtle issues for future investigation. The issue may or may not turn out to be orthogonal to our present working hypothesis, according to which consequent-imperativization is a CoP-neutral manipulation.

2.2 Polar Questions Disfavor CoP

In the previous section, the 'mood switch' in terms of consequent-imperativization was hypothesized to be CoP-neutral. But an 'interrogative switch' to a polar question has been observed not to be [11,14,29].[9] [11], who discusses French examples, observes that the necessity-component of CoP does not project out of polar questions like a presupposition would:

(14) Est-ce que, si Pierre vient, Jacques partira?
is-this that if Pierre comes Jacques leave-FUT
'If Pierre comes, will Jacques leave?' cf. [11]:35
⇝̸ If Pierre doesn't come, Jacques will stay.

In other words, polar questions are potentially CoP-disfavoring. In this section, we confirm this for both language groups under consideration. With [14], we think the QUD-approach can account for this observation. But we also turn to a more intricate case [11] adduces as an argument in favor of his claim; this involves a 'counterfactual' variant of (14).

Based on Ducrot's observation in (14), we get the following pattern. While we (tend to) have CoP in the unembedded case (15-a), it disappears in the interrogative case (15-b).

---

[8] In (13-a), the position of the *if*-clause has been changed, compared to (12-a), in order to better fit the respective QUD. We will return to this matter in Sect. 3.1.

[9] [14] credits [19] for the very same observation.

(15) **Interrogative CoP-cancellation** ([11])

a. if p, q $\rightsquigarrow_{\text{CoP}}$ if $\neg$p, $\neg$q
b. if p, q? $\not\rightsquigarrow_{\text{CoP}}$ if $\neg$p, $\neg$q

Does this pattern persist in Chinese and English alike? We think so: neither the conditional question in (16-a) nor its Chinese counterpart in (16-b) are perfected.

(16) a. If Jerry comes, will Elaine go?
b. Yaoshi Jerry lai, Elaine jiu qu ma?
if Jerry come Elaine then go Q
$\not\rightsquigarrow$ If Jerry doesn't come, Elaine will stay.

[14] argues that the QUD-approach can handle such cases: one's primary interest for asking *if p, q?* is to know whether q follows from p, and not whether q follows from something *other than* p. This additional information would be 'out of the question', so to speak. It even seems right to us to say that such questions openly instantiate the CoP-neutral QUD2 *what if p?*, a difference being that *if p, q?* constitutes a more specific variant in that it exposes no interest whatsoever in knowing if anything *other than q* follows from p. Given (16), this pragmatic explanation may apply to both Chinese and E/G.[10]

In sum, we have identified polar questions as a CoP-disfavoring factor in both Chinese and E/G. Having identified one CoP-*neutral* and one CoP-*dis*favoring factor shared by both Chinese and E/G, we now turn to a factor that can be taken to *favor* CoP in both groups of languages: the formation of *only if* conditionals.

---

[10] To substantiate his claim that polar questions are what we call a CoP-disfavoring environment, [11] also discusses what [16] would call an 'X-marked' variant of the conditional question in (14), with 'X-marking' standing for what used to be called 'counterfactual' or 'subjunctive' marking:

(i) Est-ce que, si Pierre était venu, Jacques serait parti?
is-this that if Pierre be.X come Jacques be.X left
'If Pierre had come, would Jacques have left? cf. [11]:40
$\not\rightsquigarrow_{\text{CoP}}$ Jacques didn't leave.

X-marked conditionals typically receive counterfactual interpretations, implying both their antecedents and their consequents to be false, yet both implications have been claimed to be defeasible in the literature; see [2,16,31,53], among many others. Crucially, [11] claims (i) <u>not</u> to imply that Jacques didn't leave, i.e. <u>not</u> to imply its consequent to be false. For reasons that space restrictions keep us from further elaborating on, he ascribes the non-falsity of the consequent to an absence of CoP; see also [34], who draws the very same connection between consequent-falsity and CoP. It remains to be seen whether this carries over to Chinese, which does have expressive strategies of favoring counterfactual inferences, in spite of lacking verbal X-marking [33]. We should also point out, however, that CoP-inferences might be weaker in X-marked conditionals to begin with, a concern raised by a reviewer as well as in [29,51], the latter of which provide experimental data supporting this concern.

2.3 *Only if* Favors CoP

A CoP-favoring factor *par excellence* is the formation of an *only if* conditional: after all, *only if* readily offers itself to paraphrase CoP-readings. One might even claim CoP to be no longer pragmatic, but semantic here, i.e. *entailed* rather than *implicated*, and this is what seems to be taken for granted by both [3,56]. Contrary to that, [24,25] argues that while *only if p, q* does entail p's *necessity* for q, it is not obvious that it still entails p's *sufficiency* for q; see [42] for empirical evidence. For CoP, we need both sufficiency and necessity, the former of which is contributed by bare *if*: ***if and** only if*, or *iff* for short. In other words, there is reason to believe that CoP remains an implicature under *only if*, with the difference that what is implicated is now *sufficiency*, not *necessity.*

How are *only if* conditionals expressed in Chinese? We will focus on what we call 'bare *cai*-conditionals' here: conditionals lacking an overt conditional connective and featuring only the exclusive particle *cai* 'only' in their consequents. Bare *cai*-conditionals boil down to the form *p* CAI *q*, where *cai* belongs to the consequent clause. It has been described as conveying p's necessity for q in this position [7,38]; obvious syntactic differences aside, its interpretive effect is thus similar, if not identical to, *only* in *only if.*[11]

[23] shows that *only if p, q* entails p to be *necessary*, but not also *sufficient*, for q. The same can be shown to hold of bare *cai*-conditionals in Chinese. The necessity-component of *only if* may be unsurprising. The infelicity of the following sequence of sentences, inspired by similar examples in [24], lends support to the view that *only if* entails rather than implicates this component; the second sentence is an infelicitous attempt at denying that hard work is necessary for success:

(17) Only if you work hard do you succeed, # though even if you don't work hard, you might still succeed. cf. [24]:(20)

The necessity-implication of bare *cai*-conditionals seems equally uncancelable, in support of the same conclusion:[12]

---

[11] We leave aside conditionals whose antecedents are introduced by the exclusive particle *zhi-you* 'only' (lit. 'only-have'), and are of the form *zhi-you p* *(CAI) *q* [26]. One reason for this limitation is the obligatory presence of *cai*, which already suffices to induce a necessity-reading. [58] argues *zhi-you*-conditionals to pattern with *only if* conditionals by virtue of coming with necessity, but lacking sufficiency. For the complex relationship between *zhi-you* and *cai*, see [27,28,55].

[12] The examples cited from [58] in (18) and (21) differ from their originals in two respects: (i) they lack the conditional connective *zhi-you* 'only-have', for the reasons given in footnote (see footnote 11); (ii) their consequent clauses contain the future modal *hui* 'will', instead of the possibility modal *neng* 'can'. This substitution served to ensure that the implicational properties argued for on the basis of (18) and (21) weren't due to *neng*'s existential force.

(18) Ni hen nuli cai hui chenggong, #danshi ni bu nuli ye
you very diligent CAI will succeed #but you not diligent also
keneng chenggong.
maybe succeed
'Only if you're diligent can you succeed; #but if you're not diligent, you might also succeed.'

cf. [58]:(9)+(10b)

The sufficiency-implication, by contrast, proves to be cancelable. p's semantic non-sufficiency for q is witnessed by the felicity of the second sentence in (19), which should be infelicitous if the preceding *only if* conditional entailed hard work to be sufficient for success:[13]

(19) You succeed **only if** you work hard. But sometimes when you work hard you don't succeed. [24]:(20), emphases added

The sufficiency-denying continuation in (19) is shown by [24] to become odd after an *iff*-conditional, which is thus shown to entail sufficiency (on top of necessity), and hence to semantically encode biconditionality:

(20) You succeed **if and only if** you work hard. #But sometimes when you work hard you don't succeed. [24]:(20), emphases added

So adding *only* to *if* semantically adds necessity, but apparently comes at the cost of sufficiency; see also experimental work by [42] on German. This insufficiency-pattern is again replicable with bare *cai*-conditionals:[14]

[13] However, [5] show in a footnote that sufficiency puzzlingly arises with *only if* under negation:

(i) Not only if you work hard do you succeed, #though even if you work hard, you might still fail.

cf. [5]: footnote 18, ex. (iii)

[14] Full CoP seems to require something in addition, such as the necessity adverb *yiding* 'definitely' in the consequent, an observation [58] owes to an anonymous reviewer:

(i) Ni hen nuli cai **yiding** hui chenggong; #dangran nuli-le ye
you very diligent CAI **definitely** will succeed #of.course diligent-ASP also
you keneng bu chenggong.
have possibility not succeed
'Only if you're diligent will you definitely succeed; #but of course, (even) if you're diligent, you might still fail.'

cf. [58]:(100)

A compositional analysis of such examples is not immediately obvious.

(21) Ni nuli cai hui chenggong, danshi ni jishi nuli ye
you diligent CAI will succeed but you even.if diligent also
keneng bu chenggong.
maybe not succeed
'Only if you are diligent can you succeed; but even if you're diligent, you might still not succeed.' cf. [58]:(9)+(10a)

In short, both *only if* and *cai* encode necessity, but neither of them encodes sufficiency. If these observations are correct, sufficiency is merely implicated by *only if* and *cai*, and we have reason to believe that both merely *favor*, rather than *encode*, biconditionality: they remain as semantically 'imperfect' as bare *if*-conditionals are known to be. The picture emerges that the two types of conditionals complement each other in virtue of entailing and implicating what the other does not entail or implicate:

	entailed	**implicated**
if p, q	sufficiency	necessity: 'only if p, q'
only if p, q	necessity	sufficiency: 'if p, q'

An empirical question to be addressed in future work is whether both constructions are 'imperfect' *to the same extent.* This is a pragmatic question, not a semantic one. A CoP-implicature for bare *if*-conditionals is one of *necessity*; a CoP-implicature for *only if* conditionals is one of *sufficiency.* The strength of the respective implicature will determine the strength of CoP in each case. If the sufficiency-implicature of *only if* turns out to be stronger than the necessity-implicature of bare *if*, our classification of *only if* as CoP-favoring can be maintained. If it turns out to be only equally strong or even weaker, this would constitute a reason to classify it as CoP-neutral, as counterintuitive as this may seem. And it of course cannot be taken for granted that the CoP-inferences of *only if* and *cai* are equally strong.

2.4 Section Summary

In this section, we considered three manipulations to a conditional C of the form *if p, q*, all of which seem equally (ir)relevant in both Chinese and E/G for whether C is perfected or not. Table 2 below, to be completed in this paper's conclusion, summarizes our preliminary findings:

First, we found the 'imperativization' of the consequent q to be a *CoP-neutral* manipulation. Second, the formation of conditional questions was found to clearly disfavor CoP, in line with previous observations. Finally, *only if* and its Chinese counterpart *cai* were both found to be CoP-favoring, with the caveat that this hinges on the strength of a perfecting sufficiency-implicature. Having spotlighted these interlinguistic *similarities*, we now turn to two *differences*: CoP-favoring manipulations that are available only in E/G, but not in Chinese.

Table 2. CoP-inferences in English/German vs. Chinese under certain grammatical manipulations; baseline: *if p, q*

	English/German	**Chinese**
if p, q-IMP	//	//
if p, q?	---	---
***only** if*	(+)	(+)

// = CoP-neutral

--- = CoP-disfavoring

+ = CoP-favoring

3 Differences

In this section, we review two CoP-favoring factors that are (partly) available in E/G, but not in Chinese: the position of the *if*-clause (left vs. right), and focus on conditional *then*. An *if*-clause's right-adjunction and focus on *then* can both be said to be CoP-favoring manipulations. However, neither of the two seems available in Chinese, whose grammar proves to be more restrictive in these aspects.

3.1 Left vs. Right *if*-clauses

In E/G, the **position of the *if*-clause** (left vs. right) may vary across discourse contexts [13]. On our intuition, right-adjoined antecedents favor CoP more than left-adjoined antecedents do, though we admit that this is (again) a rather subtle matter. (22-b) seems to slightly more strongly suggest than (22-a) that mowing the lawn is the only thing the addressee can do to get $5 from the speaker:

(22) a. If you mow the lawn, I'll give you $5. [if p] q
b. I'll give you $5 if you mow the lawn. q [if p]

This effect is not unexpected under information-structural accounts splitting a given sentence into an initial *topic* and a subsequent *focus*, which we take to be roughly equivalent to the notions 'given' and 'new'.[15] Under such a view, the structural difference exemplified in (22) comes with a difference in focus structure. Right-adjoined antecedents like in (22-b) tend to be focused. On the 'rule of thumb' mentioned in the introduction, intonational focus on the antecedent ($_p$ you mow the lawn) correlates with a CoP-favoring QUD1 asking about the antecedent(s) verifying the consequent ($_q$ you get $5). By contrast, left-adjoined antecedents like in (22-a) tend to have focus on the consequent, hence be more readily interpreted as answers to a CoP-neutral QUD2 asking for the consequent(s) following from the antecedent. We are not claiming that this tendency

[15] This may be controversial e.g. in light of the existence of *contrastive topics*, which there is a substantial amount of work on.

cannot be overridden, i.e. that CoP cannot be favored by focusing some part of the left-adjoined antecedent in (22-a), or being disfavored by focusing some part of the 'topicalized' consequent in (22-b). Still, we take these focus patterns to run counter to a default preference one would probably have in a forced-choice kind of setting.

The view that the position for the antecedent matters for whether or not CoP arises has been entertained before. [4] ascribes such a view to [8], who deals with the effects of putting a 'modifier' before vs. after the 'head', and treats CoP as a subphenomenon of a more general pattern that might nowadays be referred to as *contrastive focus*.[16] [8] discusses the following pair of conditionals:

(23) a. If you come, I'll help you.
b. **I'll help you if you come.**

[8]:1126, emphasis added

According to [8], "the second [conditional in (23)] lends itself more readily to the implication **'I'll help you ONLY if you come'**." ([8]; emphasis and bracketed material added). About cases he treats as parallel, he even observes that what comes second (appears to the right) is what is being *asked* about. This connects with the QUD-approach to CoP before it even existed: *antecedents* to the right answer the CoP-favoring QUD1 after all.

In a critical discussion of [22]'s view of *if*-clauses as *topics*, [13] voices the intuition that under what we call a CoP-favoring QUD1, it is somewhat odd *not* to right-adjoin the antecedent. Focus marking required by the QUD cannot override this preference:[17]

(24) Under what conditions will you buy this house?
a. I'll buy this house if you give me the money$_F$.
b. #If you give me the money$_F$, I'll buy this house.

cf. [18], via [13]:81

However, right-adjunction of the antecedent may remain only a *necessary* condition for CoP to arise: under what we call a CP-neutral QUD2 here, both sequences are judged to be acceptable by [13], at least as long as focus is on the consequent:

(25) What will you do if I give you the money?
a. I'll buy this house$_F$ if you give me the money.
b. If you give me the money, I'll buy this house$_F$.

cf. [18], via [13]:81

[16] That linearization matters for CoP appears to also be lingering in the background of [20]'s discussion of exclusive vs. nonexclusive readings for *only* in conditionals of the form *if only p, q*. The exclusive reading is observed by [20] to not arise as easily with 'left' as with 'right' antecedents. [57] ascribes the exclusive reading to the presence of CoP, and the non-exclusive reading to its absence.

[17] [13] ascribes a variant of (24) to [18]; the F-subscripts were added by us.

In Chinese, an intended CoP-reading does not require the antecedent to appear to the right of the consequent – in fact, such a configuration does not even seem possible ([46], [43]; Muyi Yang, pc):[18]

(26) ??Women qu sanbu, ruguo tianqi hao.
??we go walk if weather good
(intended:) 'We'll go for a walk if the weather is fine.'

Chinese focus and sentence structure hence do not correlate in the same way as in E/G, i.e. focus on an antecedent cannot preferably coincide with its right-adjunction. More concretely, something in the grammar of Chinese blocks the CoP-favoring strategy of right-adjoining the antecedent. As usual, however, things are not quite as simple as they may seem. For one thing, the *if*-clause can occur between the subject NP and the VP of the consequent:

(27) Women [ruguo tianqi hao] qu sanbu.
we [if weather good] go walk
'We'll go for a walk if the weather is fine.'

However, we still take the *if*-clause to be left-adjoined in (27), with the subject NP having moved out of the consequent for purposes such as topicalization (Ming Xiang, pc). At the same time, sentences like (27) do come a bit closer to the CoP-favoring configuration involving right-adjoined antecedents in E/G; it remains to be seen whether they favor CoP to a higher degree than minimally different variants where no subject movement has taken place.

What further weakens the Chinese ban on right-adjoined antecedents is that not all such cases are equally bad: as [46] observe, the acceptability may vary with different 'then'-particles inserted into the consequent. The particle *jiu* (lit.: 'already') cannot fully rescue the sentence, but is a better choice than *name*, which induces plain ungrammaticality:[19]

(28) Women {?**jiu** / ***name**} qu sanbu, ruguo tianqi hao.
we {?JIU / *NAME} go walk if weather good

[18] As [46] observe, this preference for 'left' antecedents counts as an argument in favor of [22]'s view that *if*-clauses *are* (or tend to be) topics in Chinese.

[19] As Johan van der Auwera (pc) points out to us, English *then* comes with something close to a ban on such cataphoric uses as well; cf. also [6,50]:

(i) You (*then) get (*then) a reward, if you mow the lawn.

[50] argues that this ban holds only so long as *then c-commands* the antecedent, which is not the case in the following example he provides:

(ii) Because I would then hear lots of people playing on the beach, I would be unhappy if it were sunny right now. cf. [50]:(56c)

By contrast, [46] observe *na* (lit. 'that'), the first component of *name*, to be acceptable as a cataphoric anticipation of a right-adjoined antecedent:

(29) (Na) women qu sanbu, ruguo tianqi hao dehua.
(then) we go walk if weather good COND
'We will go for a walk, if the weather is good.' cf. [46]:(41)

The example in (29) also suggests there to be acceptable cases with antecedents to the right after all. However, there seems to be an intonational break after the consequent, which sets them apart from (some) of their E/G-counterparts, and aligns with the view that 'right' antecedents serve to express an *afterthought* in Chinese, a concept [46] ascribe to earlier work. If there is some connection between such afterthoughts and CoP, it remains to be worked out. But they don't strike us as closely connected to CoP as 'right' antecedents are under the QUD-approach to CoP.

In sum, then, putting the antecedent to the right of the consequent is a more viable CoP-favoring strategy in E/G than it is in Chinese, where this configuration is only marginally available. That being said, we turn to a second CoP-favoring strategy Chinese is excluded from: focus on conditional *then*.

3.2 Focus on Conditional *then*

[50] observes focus on conditional *then* to have a CoP-like effect. This can be illustrated with the following pair of examples from German, where focused *then* is used cataphorically in one case and anaphorically in the other:[20]

(30) a. Du wirst dann$_F$ belohnt, wenn du den Rasen mähst.
you get then$_F$ rewarded if you the lawn mow
b. Wenn du den Rasen mähst, dann$_F$ wirst du belohnt.
if you the lawn mow then$_F$ get you rewarded
⇝ if you do something other than mowing the lawn, you will not get rewarded

In Chinese, stress on the 'then'-particle *jiu* comes out as plainly ungrammatical in parallel examples:[21]

(31) a. *Women jiu$_F$ qu sanbu, ruguo tianqi hao.
*we then$_F$ go walk if weather good
b. *Ruguo tianqi hao, women jiu$_F$ qu sanbu.
*if weather good we then$_F$ go walk

[20] There seem to be subtle differences in interpretation between these examples. It is fair to assume, however, that some variant of CoP is involved in both cases; see [60] for further discussion.

[21] As Muyi Yang (pc) points out, *jiu* might carry more intonational weight in the consequent of conditionals involving the conditional complementizer *zhi-yao*, lit. 'only-need'. According to the view held in [58], this kind of conditional has no CoP-reading, but a *minimal sufficiency* reading in the sense of [20].

Why are these cases bad? A first answer comes from generalizations from the existing literature on *jiu* [1,26,39,41,43]: for one thing, 'cataphoric' conditional *jiu* is marked to begin with, as seen in the preceding subsection. Descriptively speaking, there are at least two *jiu*-variants, a left-associating jiu_{i} and a right-associating jiu_{ii}, and conditional *jiu* is of the former kind. More crucially, jiu_{i} is always unfocused, as opposed to jiu_{ii}, which is always focused.

The two *jius* also differ in meaning: in a nutshell, while jiu_{ii} is exclusive and means 'only', jiu_{i} is nonexclusive and often means 'already' [38]). Illustrating the difference between the two, [41] provides the following pair of examples:

(32) a. Yuehan_{F} jiu_{i} hui shuo fayu.
John_{F} JIU_{i} can speak French
'John can speak French.'

b. Jiu_{ii} Yuehan_{F} hui shuo fayu.
JIU_{ii} John_{F} can speak French
'**Only** John can speak French.' cf. [41]

Given uses like (32-b), one may wonder why (31-a) doesn't have the same CoP-reading as the German examples involving focused 'then' in (30): given that carrying focus distinguishes exclusive jiu_{ii} from nonexclusive jiu_{i}, focus on *jiu* could be expected to induce a jiu_{ii}-reading, and the antecedent to its right could serve as its focus associate.[22] But rather than improving an already marked configuration, focus on *jiu* makes things worse.[23]

In this brief subsection, only the focused variant of conditional *then* was considered. But it has been claimed in previous work that even unfocused *then* comes with some form of CoP [13,30,32]: *if p, then q* is taken to convey that at least *some* cases in which p is false are cases in which q is also false. We call this a *weakly exclusive* CoP-reading here.[24] [26] entertains a parallel view of Chinese *jiu*, proposing a weakly exclusive *semantics* for it. [41] objects that this weak exclusiveness is merely an *implicature*, and even proposes that structures involving nonexclusive jiu_{i} are 'anti-exhaustive', to the exclusion of strong CoP. So not even unfocused 'then' seems to favor CoP in Chinese, at least not to the same extent as it does in E/G.

[22] We thank Muyi Yang (pc) for discussion on this point. [40] does treat conditional *jiu* as an 'only'-like jiu_{ii} whose focus associate is the antecedent, and explains its apparent nonexclusiveness with the set of focus alternatives, which are such that jiu_{ii} cannot exclude anything from them; [41] pursues the same strategy for the jiu_{i} seen in (32-a).

[23] A syntactic explanation for the badness of (31-a) could be that focused *jiu* must be adjacent to its focus. Even if we take the entire antecedent to be focused in (31-a), *jiu* is still not adjacent to it.

[24] [32] entertains what we call a *strongly* exclusive CoP-reading: *all* $\neg$p-cases are implied to be $\neg$q-cases.

3.3 Section Summary

In this section, we looked at two CoP-favoring factors in E/G which seem partially blocked by the grammar of Chinese: right-adjoined antecedents and focus on conditional *then*. We did not come up with any factors that favor CoP in Chinese, but not in E/G. It would be too early to conclude that no such factors exist, but if they didn't, it would have to be explained why Chinese should have less CoP-favoring factors at its disposal: does context play a more crucial role in inducing CoP-readings? Or are the available favoring factors, like the *cai*-insertion into the consequent considered in this section, simply more 'powerful' in terms of the graded notion of CoP we are entertaining?

4 Conclusion

This paper's main goal was to take a step towards widening the crosslinguistic picture of conditional perfection (CoP), identifying CoP-favoring vs. -disfavoring factors across two typologically distinct groups of languages: Chinese on the one hand and English/German [E/G] on the other. At a more general level, such an endeavor has broader implications for the crosslinguistic emergence of so-called *exhaustivity* implicatures, which CoP clearly falls under. Put in psycholinguistic terms, our procedure was to treat CoP as a *dependent* variable and see how it behaves if we manipulate several *independent* ones. These were divided into 'language group' (Chinese vs. E/G) and certain manipulations. While we classified all of these manipulations as 'grammatical', we took each of them to have a pragmatic dimension, assuming [14]'s QUD-approach to CoP.

The grammatical manipulations and their language-specific effects are summed up in Table 3 below. The baseline construction subjected to these manipulations was always a bare (indicative) conditional of the form *if p, q*. We distinguished between three kinds of factors: CoP-favoring, CoP-*dis*favoring, and CoP-neutral. The first three rows indicate a certain uniformity across the two groups under comparison. The table's last two rows reveal clear interlinguistic differences: two factors tentatively seen as CoP-favoring in E/G – right-adjoined antecedents and focus on conditional *then* – are grammatically blocked in Chinese.

More research will be needed to both broaden and deepen the empirical picture, and provide formally explicit analyses. There is a lot of subtlety involved, calling for a clarification of the empirical picture via corpora or experimentation. Another potentially CoP-relevant factor to be considered is the type of text in which a conditional appears. A reviewer raises the question of whether legal texts count as CoP-favoring environments, for example. This may again be subject to crosslinguistic variation. Relatedly, many more languages will have to be taken into account to come closer to true crosslinguistic universals. At the same time, we hope to have shown how crosslinguistic contrasts like the ones sketched in this paper can inform, and maybe lead to a refinement of, existing theories of CoP, and of exhaustivity implicatures in general.

Table 3. CoP-inferences in English/German vs. Chinese under certain grammatical manipulations; baseline: *if p, q*

	English/German	**Chinese**
if p, q-IMP	//	//
if p, q?	---	---
only if	(+)	(+)
q if p	+	?*
conditional *then*$_F$	+	?*

// = CoP-neutral
--- = CoP-disfavoring
\+ = CoP-favoring

Acknowledgement. For valuable comments, we are indebted, among others, to our reviewers for TLLM 4 in Beijing, Sinn und Bedeutung (SuB) 28 in Bochum and the 10th International Contrastive Linguistics Conference (ICLC 10) in Mannheim, the audiences at these events, as well as Johan van der Auwera, Johanna David, Daniel Hole, Magdalena Kaufmann, Małgorzata Kielak, Dean McHugh, Robert van Rooij, Stephanie Rotter, Marvin Schmitt, Katrin Schulz, Frank Sode, Yen Quynh Vu, Danfeng Wu, Ming Xiang, Muyi Yang and Xuetong Yuan. All remaining shortcomings are our own.

Disclosure of Interests. The authors have no competing interests to declare. A.W. and M.L. jointly conceived and designed the research reported in this paper. Both authors were responsible for the interpretation of the data, which include M.L.'s judgments on the Chinese examples. A.W. prepared the first draft of the manuscript; both authors contributed to its revision. M.L. provided funding acquisition, project administration, and resources.

References

1. Alleton, V.: Les adverbes en chinois moderne. De Gruyter (1972)
2. Anderson, A.R.: A note on subjunctive and counterfactual conditionals. Analysis **12**, 35–38 (1951)
3. van der Auwera, J.: *Only if*. Logique et analyse **28**(109), 61–74 (1985)
4. van der Auwera, J.: Pragmatics in the last quarter century: the case of conditional perfection. J. Pragmat. **27**(3), 261–274 (1997)
5. Bassi, I., Bar-Lev, M.E.: A unified existential semantics for bare conditionals. In: Truswell, R., Cummins, C., Heycock, C., Rabern, B., Rohde, H. (eds.) Proceedings of Sinn und Bedeutung (SuB) 21, pp. 125–142 (2018)
6. Bhatt, R., Pancheva, R.: Conditionals. In: Everaert, M., van Riemsdijk, H.C. (eds.) The Blackwell companion to syntax, pp. 1–48. Blackwell, 2 edn. (2017). https://doi.org/10.1002/9781118358733.wbsyncom119
7. Biq, Y.O.: The semantics and pragmatics of *cai* and *jiu* in Mandarin Chinese. Ph.D. thesis, Cornell University (1984)

8. Bolinger, D.L.: Linear modification. PMLA **67**(7), 1117–1144 (1952)
9. Cariani, F., Rips, L.J.: Experimenting with (conditional) perfection: tests of the exhaustivity theory. In: Kaufmann, S., Over, D.E., Sharma, G. (eds.) Conditionals: logic, linguistics and psychology, pp. 235–274. Palgrave Macmillan (2023). https://doi.org/10.1007/978-3-031-05682-6_9
10. Condoravdi, C., Lauer, S.: Conditional imperatives and endorsement. In: Lamont, A., Tetzloff, K. (eds.) Proceedings of NELS 47, pp. 185–204 (2017)
11. Ducrot, O.: Présupposés et sous-entendus. Lang. Fr. **4**, 30–43 (1969)
12. Farr, M.C.: Focus influences the presence of conditional perfection: experimental evidence. In: Reich, I., Horch, E., Pauly, D. (eds.) Proceedings of Sinn und Bedeutung (SuB) 15, pp. 225–240 (2011)
13. von Fintel, K.: Restrictions on quantifier domains. Ph.D. thesis, UMass Amherst (1994)
14. von Fintel, K.: Conditional strengthening: a case study in implicature (2001), ms., MIT
15. von Fintel, K., Iatridou, S.: A modest proposal for the meaning of imperatives. In: Arregui, A., Rivero, M.L., Salanova, A. (eds.) Modality across syntactic categories, pp. 288–319. OUP, Oxford (2017). https://doi.org/10.1093/acprof:oso/9780198718208.003.0013
16. von Fintel, K., Iatridou, S.: Prolegomena to a theory of X-marking. Linguist. Philos. **46**, 1467–1510 (2023). https://doi.org/10.1007/s10988-023-09390-5
17. Geis, M.L., Zwicky, A.M.: On invited inferences. Linguistic Inquiry **2**(4), 561–566 (1971)
18. Givón, T.: Logic vs. pragmatics, with human language as the referee: toward an empirically viable epistemology. J. Pragmatics **6**, 81–133 (1982)
19. Groenendijk, J.A.G., Stokhof, M.J.B.: Studies on the semantics of questions and the pragmatics of answers. Ph.D. thesis, Universiteit van Amsterdam (1984)
20. Grosz, P.: On the grammar of optative constructions. John Benjamins (2012)
21. Grusdt, B., Liu, M., Franke, M.: Testing the influence of QUDs on the occurrence of conditional perfection. In: Knowlton, T., Schwarz, F., Papafragou, A. (eds.) Experiments in Linguistic Meaning (ELM) 2, pp. 104–116 (2023)
22. Haiman, J.: Conditionals are topics. Language **54**, 564–589 (1978)
23. Herburger, E.: Conditional perfection: the truth and the whole truth. In: D'Antonio, S., Moroney, M., Little, C.R. (eds.) Proceedings of SALT 25, pp. 615–635 (2015)
24. Herburger, E.: *Only if*: if only we understood it. In: Csipak, E., Zeijlstra, H. (eds.) Proceedings of Sinn und Bedeutung (SuB) 19, pp. 304–321 (2015)
25. Herburger, E.: Bare conditionals in the red. Linguist. Philos. **42**(2), 131–175 (2019). https://doi.org/10.1007/s10988-018-9242-2
26. Hole, D.: Focus and background marking in Mandarin Chinese: system and theory behind *cái*, *jiù*, *dōu* and *yě*. Routledge (2004)
27. Hole, D.: A crosslinguistic syntax of scalar and non-scalar focus particle sentences: the view from Vietnamese and Chinese. J. East Asian Linguis. **26**(4), 389–409 (2017). https://doi.org/10.1007/s10831-017-9160-2
28. Hole, D.: The relationship between Chinese *zhiyou* 'only' and *cai*: a matter of morphosyntax. J. East Asian Linguist. **32**, 1–16 (2023). https://doi.org/10.1007/s10831-023-09257-7
29. Horn, L.R.: From *if* to *iff*: conditional perfection as pragmatic strengthening. J. Pragmat. **32**(3), 289–326 (2000)
30. Iatridou, S.: On the contribution of conditional *then*. Nat. Lang. Seman. **2**, 171–199 (1993)

31. Iatridou, S.: The grammatical ingredients of counterfactuality. Linguistic Inquiry **31**, 231–270 (2000)
32. Izvorski, R.: The syntax and semantics of correlative proforms. In: Kusumoto, K. (ed.) Proceedings of NELS 26 (1996)
33. Jiang, Y.: Ways of expressing counterfactual conditionals in Mandarin Chinese. Linguist. Vanguard **5** (2019). https://doi.org/10.1515/lingvan-2019-0009
34. Karttunen, L.: Counterfactual conditionals. Linguistic Inquiry **2**, 566–569 (1971)
35. Kaufmann, M.: Interpreting imperatives. Springer (2012)
36. Kaufmann, M.: Imperative clauses. In: Guesser, S., Marchesan, A., Junior, P.M. (eds.) Wh-exclamatives, imperatives and wh-questions, pp. 173–225. De Gruyter (2023). https://doi.org/10.1515/9783111183176-007
37. Kaufmann, S., Schwager, M.: A unified analysis of conditional imperatives. In: Cormany, E., Ito, S., Lutz, D. (eds.) Proceedings of SALT 19, pp. 239–256 (2009)
38. Lai, H.L.: Rejected expectations: the scalar particles *cai* and *jiu* in Mandarin Chinese. Linguistics **37**, 625–661 (1999)
39. Liu, M.: Exclusive and non-exclusive ONLYs in Chinese and English (2016), ms. for WCCFL 34
40. Liu, M.: Mandarin conditional conjunctions and *only*. Stud. Logic **10**, 45–61 (2017)
41. Liu, M.: Varieties of alternatives: Mandarin focus particles. Linguist. Philos. **40**, 61–95 (2017). https://doi.org/10.1007/s10988-016-9199-y
42. Liu, M., Barthel, M.: Semantic processing of conditional connectives: German *wenn* 'if' versus *nur wenn* 'only if'. J. Psycholinguistic Res. **50**, 1337–1368 (2021). https://doi.org/10.1007/s10936-021-09812-0
43. Liu, M., Wang, Y.: *Jiu*-conditionals in Mandarin Chinese: thoughts on a uniform pragmatic analysis of Mandarin conditional constructions. Linguistics Vanguard **8**, 435–446 (2022). https://doi.org/10.1515/lingvan-2021-0036
44. McCawley, J.: Everything that linguists have always wanted to know about logic* *but were ashamed to ask. University of Chicago Press (1993)
45. Oikonomou, D.: Covert modals in root contexts. Ph.D. thesis, MIT (2016)
46. Pan, V.J., Paul, W.: The syntax of complex sentences in Mandarin Chinese: a comprehensive overview with analyses. Linguistic Analysis **42**, 63–162 (2018)
47. Parsons, J.: Imperative conditionals (2015), ms. https://philpapers.org/archive/PARIC.pdf
48. Roberts, C.: Information structure: towards an integrated formal theory of pragmatics. Semantics and Pragmatics **5**, 1–69 (2012). https://doi.org/10.3765/sp.5.6, based on a working paper from 1996
49. Rooth, M.: A theory of focus interpretation. Nat. Lang. Seman. **1**, 75–116 (1992)
50. Schlenker, P.: Conditionals as definite descriptions. Res. Lang. Comput. **2**, 417–462 (2004)
51. Schwab, J., Liu, M.: Processing attenuating NPIs in indicative and counterfactual conditionals. Front. Psychol. **13** (2022). https://doi.org/10.3389/fpsyg.2022.894396
52. Schwager, M.: Conditionalized imperatives. In: Gibson, M., Howell, J. (eds.) Proceedings of SALT 16, pp. 241–258 (2006)
53. Stalnaker, R.: Indicative conditionals. Philosophia **5**, 269–86 (1975)
54. Stegovec, A.: Perspectival control and obviation in directive clauses. Nat. Lang. Seman. **27**(1), 47–94 (2019). https://doi.org/10.1007/s11050-019-09150-x
55. Sun, Y.: A bipartite analysis of *zhiyou* 'only' in Mandarin Chinese. J. East Asian Linguist. **30**, 1–37 (2021). https://doi.org/10.1007/s10831-021-09228-w

56. Van Canegem-Ardijns, I., Van Belle, W.: Conditionals and types of conditional perfection. J. Pragmat. **40**(2), 349–376 (2008). https://doi.org/10.1016/j.pragma.2006.11.007
57. Wimmer, A.: Keeping *only* exclusive in conditional antecedents. In: Aitha, A., Castro, S., Wilson, B. (eds.) Proceedings of CLS 57, pp. 445–459 (2022)
58. Wimmer, A.: *Zhi-*{*yao, you*} 'only-need, have': on two conditional connectives in Mandarin. J. East Asian Linguist. **31**, 401–438 (2022). https://doi.org/10.1007/s10831-022-09243-5
59. Wimmer, A.: Perfecting imperative conditionals. In: Phadnis, S., Spellerberg, C., Wilkinson, B. (eds.) Proceedings of NELS 54. pp. 205–214 (2024)
60. Wimmer, A., Liu, M.: Conditional *then* and the QUD-approach to conditional perfection. In: Baumann, G., Gutzmann, D., Koopman, J., Liefke, K., Renans, A., Scheffler, T. (eds.) Proceedings of Sinn und Bedeutung (SuB) 28. pp. 980–998 (2024)
61. Yuan, X.: Challenging the presuppositions of questions: the case of *ba*-interrogatives. In: Franke, M., Kompa, N., Liu, M., Mueller, J.L., Schwab, J. (eds.) Proceedings of Sinn und Bedeutung (SuB) 24, pp. 469–484 (2020)
62. Zevakhina, N.: Veridicality and the cause-effect relation in Russian *esli-* and *raz-*conditionals: experimenting with conditional perfection and logical entailment. Linguistics Vanguard **8**(s4), 401–412 (2022). https://doi.org/10.1515/lingvan-2021-0037
63. Zhu, B., Duan, J.: Types and permissible conditions of biconditional hypothetical sentences. Chinese Language Learning (Han Yu Xue Xi) **6**, 12–23 (2023)

Disjunction and Parentheses

Jialiang Yan[1], Chen Ju[2](✉), and Wei Wang[1](✉)

[1] Tsinghua University, Beijing 100084, China
weiwang.ueiwaon@gmail.com
[2] Renmin University of China, Beijing 100084, China
juchen727@163.com

Abstract. We observe that some disjunctions admit interpretations richer than the boolean interpretation by communicating intended orderings between disjuncts. Such disjunctions, which we denote as ordered disjunctions, can be encoded by disjunctions with parentheses of the form A (or B). We present a linguistic analysis of disjunctions with parentheses, and categorize their ordered interpretations into three types: preference, likelihood, and appropriateness of wording. We argue that the source of the orderings encoded by disjunctions with parentheses is pragmatic effects of parentheses weakening information inside while emphasizing information outside, derived from the basic function of parentheses to mark parentheticals. We study reasoning involving ordered disjunction, showing that ordered disjunction do not satisfy the commutative law, and it gives rise to novel types of free choice inferences and ignorance inferences. These phenomena are captured in our proposed model based on the framework of Bilateral State-based Modal Logic (BSML).

Keywords: Ordered disjunction · Parentheses · Free choice inferences

1 Introduction

Disjunctions, one of the most common logical constructions, are ubiquitous in natural language. In *classical logic*, they are represented as boolean disjunctions in the form of $A \vee B$, which is true if and only if A or B is true. Consider the following sentences:

(1) a. This room is for studying or socializing.
b. Shenti jiankang, neng shiying chuchai huo
body healthy can adapt on.business or
zhuwai gongzuo.[4] [Mandarin]
international.deployment work
'Be healthy, and able to adapt to business trips or international deployment."
c. Gyōsei kōi wa kanarazu hōritsu matawa jōrei no
administration act TOP without.fail law or ordinance GEN
sadame ni shitagat-te jisshisa-re-ru.[5] [Japanese]
rule DAT observe-GER carry.out-PASS-NPST[6]

J. Yan et al. (Eds.): TLLM 2024, LNCS 15410, pp. 127–148, 2025.
https://doi.org/10.1007/978-3-031-86054-6_7

'Administrative acts must be carried out in accordance with the provisions of laws or ordinances."

The disjunctions in the above example, when interpreted as boolean disjunctions, simply indicate that at least one of the disjuncts is true. Under the classical semantics, the semantic value of a disjunction is determined *only* by the semantic values of its disjuncts, ignoring other semantic aspects of the disjunction, such as the relationships between the two disjuncts. However, these relationships, particularly the *orderings* between disjuncts, can also be encoded by the disjunction.

For instance, suppose there is a room that is mainly intended for students to study and discuss, but on special occasions, it can also be used as a social space. In this scenario, when someone uses (1-a) to introduce the room, she might intend to convey that the room is for studying or socializing, but studying is *preferred* over socializing.

This kind of intended ordering between disjuncts can be **encoded** within the disjunction by several linguistic or orthographic devices. In *spoken* language, the speaker can employ prosodic devices, including phrasal stress or intonation emphasizing "studying." In *written* language, the order can be encoded in virtue of not only typographical means by underlining, putting the most highly ranked disjunct in boldface, italics, or another font, but also punctuation, in particular *parentheses*[4]. If the first disjunct ranks the highest according to the writer's intended ordering, in languages such as English, Chinese, and Japanese, she could put the disjunction connective and all other disjuncts in parentheses to form expressions of the form "A (or B)," to be termed as **disjunction with parentheses**. For instance:

(2) a. This room is for studying (or socializing).
b. Shenti jiankang, neng shiying chuchai (huo
body healthy can adapt on.business or
zhuwai) gongzuo. [Mandarin, [28,29]]
international.deployment work
"Be healthy, and able to adapt to business trips (or international deployment)."

---

[1] Adapted from [28,29].
[2] Adapted from [19].
[3] Abbreviations in this paper: DAT = dative, GEN = genitive, GER = gerund, NPST = non-past, PASS = passive, TOP = topic.
[4] The English word "parenthesis" has several meanings. It can mean "a word or phrase inserted as an explanation or afterthought into a passage which is grammatically complete without it, in writing usually marked off by brackets, dashes, or commas", and can also denote "a pair of round brackets () used to mark off a parenthetical word or phrase" ([25]). In this paper, we use the word "parenthesis" in the second sense, and use "parenthetical" to express the first sense of "parenthesis."

c. Gyōsei kōi wa kanarazu hōritsu (matawa jōrei)
administration act GEN without.fail law or ordinance
no sadame ni shitagat-te jisshisa-re-ru. [Japanese, [19]]
GEN rule DAT observe-GER carry.out-PASS-NPST
"Administrative acts must be carried out in accordance with the provisions of laws (or ordinances)."

In the examples, parentheses are intentionally used to encode the intended orderings between disjuncts. These orderings are part of the semantics of the disjunctions, so semantic theories failing to capture them are insufficient. An adequate semantic theory should account for the *ordered reading* of disjunctive statements, that is, the reading according to which there is an intended ordering between the disjuncts.

It is worth noting that the ordered reading discussed above differs from the ordered disjunction defined in Logic Programs with Ordered Disjunctions (LPODs, [9]), which presents an alternative interpretation of disjunction. In LPODs, an ordered disjunction $C_1 \times \cdots \times C_n$ shows a preference hierarchy: C_1 is the most preferred option, but if C_1 is impossible, C_2 is also acceptable, and so on. This *diverges* from the boolean reading of a disjunction, whereas the ordered reading of a disjunction *extends* the boolean reading with an ordering that ranks the disjuncts. We hereafter use "ordered disjunction" to refer to disjunctions admitting the ordered reading. Disjunctions with parentheses, then, is a significant sort of ordered disjunctions.

Now let us investigate if phenomena and inferences originally associated with boolean disjunctions also apply to ordered disjunctions. One key example to be discussed in this paper is the well-known phenomenon of *free choice* (see [2,15,18,30], among others), where a modal disjunction can imply a conjunctive meaning. Consider the following example:

(3) a. This room should be used for studying or socializing.
b. This room should be used for studying (or socializing).

A typical free choice inference is licensed by (3-a), i.e., both "It can be used for studying" and "It can be used for socializing" can be inferred from (3-a). However, the sentence in (3-b), which we interpret under the ordered reading, triggers a different kind of free choice, whose derivation is affected by the ordering conveyed in the sentence. Intuitively, by typical free choice effects, the implicature is that both options presented by the disjuncts can be chosen freely. However, in this case, one option is strongly suggested, so it is better to choose that one, albeit the other option is also available.

Back to the example, studying is the room's primary purpose, while socializing is a secondary function. Therefore, from (3-b), by free choice effects, it follows that "This room is primarily intended for studying" and "This room can be used for socializing." We will discuss more examples and provide a detailed account of free choice triggered by ordered disjunctions in Sect. 4.2.

Moreover, some logical laws related to boolean disjunction may no longer hold for ordered disjunction. For example, the commutative law for disjunction (the truth value of "A or B" is equal to that of "B or A") does not hold under the ordered reading, as the expression "A (or B)" imposes an ordering that ranks A higher than B. We will discuss this issue further in the paper.

The aim of this paper is to study and provide a semantics for disjunctions with parentheses. The paper is structured as follows: Sect. 2 provides a linguistic analysis of disjunctions with parentheses, classifying their ordered interpretations into three types. Section 3 analyzes the source of the orderings encoded by disjunctions with parentheses. In Sect. 4, we explore reasoning involving ordered disjunctions and in particular disjunctions with parentheses. Section 5 formalizes ordered disjunction using an extension of the Bilateral State-based Modal Logic ([2]), and show that our model predicts the phenomena observed in Sect. 4. Finally, Sect. 6 concludes the paper and presents potential directions for future research.

2 Analysis of Disjunctions with Parentheses

In this section, we present a linguistic analysis of disjunctions with parentheses, paving the way for later formalization. We classify disjunctions with parentheses into three types (Sect. 2.1), and show that the first two types share a common nature (Sect. 2.2), while the third type is of a radically different nature (Sect. 2.3).

2.1 Classifications of Disjunctions with Parentheses

We observe that disjunctions with parentheses can have an ordered reading, under which in addition to the boolean disjunctive meaning, they also convey orderings between disjuncts. Based on the types of orderings between disjuncts they convey, we categorize disjunctions with parentheses into three types.

I. *A* is preferred

In some cases, a disjunction with parentheses in the form of "A (or B)" encodes a preference ordering: both A and B are acceptable, but A is more recommended or preferred. Consider (2-a) and the following example:

(4) No matter what your normal workout schedule is, waking up 10 minutes earlier in the morning and **taking a walk around the block (or, alternatively, simply dancing to your morning radio show for 10 minutes)** will let you start off your day with a faster-burning metabolism and, most likely, head off to work in a better mood than usual. ([11])

This example is extracted from a book that instructs people how to keep fit. According to the text, the reader can choose to walk around the block or dance to the morning radio show in the morning, while the former is more recommended by the writer. The attitude of the writer is reflected in several ways: the writer used some adverbs such as "alternatively" and "simply" to imply that dancing to morning radio shows, despite being easier, is not necessarily the best one; the use of parentheses makes this implication more explicit.

This type of "A (or B)" construction where A is preferred to B can be **tested** by appending sentences of the forms "So you'd better A" and "So you'd better B." For example,

(5) a. This room is for studying (or socializing). So you'd better study here.
b. # This room is for studying (or socializing). So you'd better socialize here.

As mentioned earlier, "This room is for studying (or socializing)" admits an ordered reading, according to which studying is preferred over socializing with respect to the room's function. Under this reading, (5-a) is felicitous, while (5-b) is not. This is because it seems self-contradictory to first express "studying is the primary function of this room" and then say "you'd better socialize here."

II. *A* is more likely

Sometimes, in a disjunction with parentheses "A (or B)," A is not necessarily a better choice, but is a more likely option or a more probable scenario. Consider the following examples:

(6) You think it's morally wrong. That means you **don't eat pork (or feel guilt when you do)** and think other people shouldn't eat pork. ([11])

(7) IT salaries there are often **three times (or more)** the norm, Red Herring reported last year. ([11])

Take (6) as an example. This paragraph tries to draw some conclusions from believing that something is morally wrong. When drawing conclusions from a fact, a rational person should order the conclusions by their objective likelihood from her point of view, rather than her subjective preference. Assume that a person believes that something is immoral, the most likely scenario is that she avoids doing it, and the second most likely scenario is that she does it, but will feel guilt afterwards.

Given a sentence with an "A (or B)" construction, we can **test** whether this construction is an instance of type II by appending sentences of the forms "It is more likely that A" and "It is more likely that B." For instance,

(8) IT salaries there are often **three times (or more)** the norm. It is more likely that they are often three times the norm.

(9) #IT salaries there are often **three times (or more)** the norm. It is more likely that they are often five times the norm.

As discussed earlier, "three times (or more)" implies that "three times" is more likely than "more." Therefore, it is felicitous to explicitly express that "three times" is more likely, but infelicitous to explicitly say that "more" is more likely.

III. *A* is more appropriate wording

In a disjunction with parentheses "*A* (or *B*)," *A* and *B* can also refer to the same thing. Some prime examples are:

(10) The point though is that **religious self-transformation (or "repentance" in Christian speak, or "illumination" in Buddhism)** is not an intellectual process. ([11])

(11) These colonies were controlled **by the English (or British)**. ([11])

(12) In addition, **such tests (or inquires)** may implicate collective bargaining rights in a unionized workplace. ([11])

In the examples, *A* and *B* are different signifiers with the same referents. "(or *B*)" is used in the text to explain and clarify the discourse, making it more complete or reader-friendly, and serving as a *metadiscourse*[5]. It tells the reader that *B* is a less precise or proper way than *A* to refer to the object in this context. In (10), the writer tries to refer to "religious self-transformation" in an objective way, without taking the perspective of any particular religion. Considering the diverse backgrounds of the readers, the writer then provides different terms used by various religions for this concept in parentheses to make the text more reader-friendly. (11) and (12) can also be understood in this manner.

Pragmatics provides motivation for the metadiscourse use. From a polyphonic perspective, the parentheses here introduce possible voices – some other ways to say the same thing. The term **polyphony** here refers to a linguistic phenomenon which "is a subtle way of bringing both self and others into a text which at first sight might be considered to be 'objective' and deprived of traces left by the author or by other voices." ([14]) By putting some disjuncts in parentheses, the writer implies that even though they are not the best or most proper expressions from *her* perspective, some of the diverse readers might still find them helpful. Thus, the writer expresses her views and introduce some others' views in the same utterance, completing a subtle interaction with the readership, which makes the discourse more reader-friendly.

This type of metadiscourse use of disjunctions with parentheses can be explicit when translated into Mandarin Chinese. In Mandarin, the metadiscourse use of "or" cannot be expressed solely by "huo (或)" or "huozhe (或者)," the Mandarin counterparts of "or." One has to juxtapose "huozhe" with "shuo (说, 'say')" to form "huozhe shuo." For instance:

[5] Metadiscourse is "the ways in which writers and speakers interact through their use of language with readers and listeners." It is "discourse about the ongoing discourse." ([17]).

(13) Zhexie zhimindi bei Yinggelan ren (huozhe shuo Buliedian
these colony PASS England people (or say British
ren) kongzhi. [Mandarin]
people) control
"These colonies were controlled by the English (or British)."

To sum up, there are three types of disjunctions with parentheses of the form "*A* (or *B*)": I. *A* is preferred, II. *A* is more likely, III. *A* is more appropriate wording.

2.2 Common Nature of Types I and II

It is noteworthy that the semantic contributions of disjunctions with parentheses of types I and II share a common nature: in both cases, an *ordering* – preferential in the case of type I, epistemic in the case of type II – is introduced, and whether the ordering is preferential or epistemic is a function of the flavor of the sentence.[6] For instance, consider again (1-a).

(1-a) This room is for studying (or socializing).

When (1-a) is perceived as being *factive*, the disjunction with parentheses gets an type I interpretation, as in our prior analysis in Sect. 1. It is also possible to understand (1-a) as expressing writer's *epistemic uncertainty* about the exact function of the room[7], which leads to an type II interpretation.

Therefore, the distinction between types I and II is not due to inherent ambiguity of disjunction with parentheses itself, but is rather derived from the sentences containing the disjunction. This underlies our universal treatment of types I and II in our formalization (see Sect. 5).

2.3 Different Nature of Type III

The three types of disjunctions with parentheses differ in whether the orderings between disjuncts are preserved under **negation**: they are lost for types I and II, but preserved for type III. This reveals that type III is of a different nature: it operates at a different – *metadiscourse* rather than discourse – level.

We explain the preservation and disappearance of orderings between disjuncts by virtue of QUD (Question Under Discussion, see [21]), an essential part of the information structure of a sentence.

I. When *A* is preferred in the affirmative counterpart
When "*A* (or *B*)" implies *A* is preferred to *B*, one may assume that "not *A* (or *B*)" would imply that *A* is not preferred to *B*. This, however, contradicts our observation:

(14) a. This room is for studying (or socializing).
b. This room is not for studying (or socializing).

[6] We thank an anonymous reviewer for pointing this out.
[7] This reading can be forced by, e.g., appending "I don't know which" to (1-a).

(15) a. You should go to see a doctor (or get medicine).
b. You shouldn't go to see a doctor (or get medicine).

(16) a. She wants to have a cup of coffee (or tea).
b. She doesn't want to have a cup of coffee (or tea).

As we observed before, the disjunction in (14-a) admits an ordered interpretation, which seems to disappear when embedded in negation, i.e., in (14-b). This is also the case for (15) and (16). In what follows, we offer a QUD-based explanation for this phenomenon.

A pragmatic theory based on QUD typically assumes that every sentence addresses a QUD either by answering it, or by raising another question that is conducive to answering the QUD. We believe that when a declarative sentence communicates an ordering, this ordering is also part of the answer to the QUD: it is an ordering over the set of possible answers to the QUD.

For instance, the QUD of (14-a) is "What is this room for?", which is an *alternative* question, whose set of possible answers is {this room is for studying, this room is for socializing, this room is for meeting, this room is for resting, ...}. The ordering encoded by "studying (or socializing)" is an ordering over this set.

According to [26], only *polar* QUDs can license a negative utterance. The QUD of (14-b) is the polar question "Is this room for studying?", whose set of possible answers is {this room is for studying, this room is not for studying}. This set does not include the second disjunct in the sentence, so the sentence does not convey any ordering between the two disjuncts.

II. When A is more likely in the affirmative counterpart

Similar to the above case, when "A (or B)" implies an ordering according to which A is more likely than B, the negation of this disjunction does not negate the ordering; in contrast, the ordering disappears. Some illustrative examples are:

(17) a. It takes hours (or days) to learn.
b. It won't take hours (or days) to learn.

(18) a. Usually you need to take 2 (or more) shots of the same pose.
b. Usually you don't need to take 2 (or more) shots of the same pose.[8]

(19) a. He lives in MUY central (or next door).
b. He doesn't live in MUY central (or next door).

[8] Some people may read from (17-b) and (18-b) the implicatures that taking days to learn is more impossible than taking hours to learn, and taking more shots is more impossible than taking 2 shots. However, this kind of implicature is not from the parentheses, but is a kind of scalar implicature. "Hours" and "days" / "2 shots" and "more shots" form a pragmatic scale (see [13]). If someone does not need to take hours to learn something, it pragmatically entails this person does not need to take days. Therefore, one may infer that taking days is more impossible. If there were no pragmatic scales, it would be difficult to get any ordered interpretation, as shown by the next example.

The reason for this phenomenon is similar to that for type I.

III. When A is more appropriate wording in the affirmative counterpart
When "A (or B)" implies the ordering according to which A is more appropriate wording, however, the ordering remains intact in "not A (or B)." For example:

(20) a. It was about religious self-transformation (or "repentance" in Christian speak).
b. It was not about religious self-transformation (or "repentance" in Christian speak).

(21) a. These colonies were controlled by the English (or British).
b. These colonies were not controlled by the English (or British).

(22) a. Most companies require tests (or inquires).
b. Most companies don't require tests (or inquires).

This can be explained by the fact that in those sentences, negation works at the discourse level, so that it cannot scope over the orderings given by disjunctions with parentheses, which are part of the metadiscourse.

3 The Source of the Orderings

In this section, we provide a linguistic explanation of the source of the ordering between disjuncts expressed by a disjunction with parentheses. We argue that the ordering comes from the *ascending and descending effects of parentheses*, i.e., pragmatic effects of parentheses weakening information in parentheses while highlighting information outside, derived from the basic function of parentheses to mark **parentheticals**.

According to discussions of the semantics and pragmatics of parentheticals (e.g., [23]), the use of parentheses is justified by the thematic deviation from the expressions to which the parentheticals are interpolated (called the "host"), which leads to semantic discontinuity. When dealing with semantic discontinuity, speakers must manage two different semantic contents within a single utterance, which is challenging given the linearity of natural language. Using parentheticals serves as a linguistic strategy to address this challenge by introducing prosodic or syntactic disruption[9], licensing the inclusion of two distinct pieces of information within one utterance. By using parentheticals, speakers put a linguistic item out of the ongoing speech act, and in this way keep it away from the addressees' primary focus. Therefore, parentheticals express *secondary information* relative to the host (see [20]).

As the basic function of parentheses is to mark parentheticals in writing, parentheses naturally inherit the pragmatic function of parentheticals to express information less important than that outside the parentheses. As a result, adding

[9] This disruption is argued by [23] to be a fundamental feature of parentheticals.

parentheses renders the information carried by the expression inside the parentheses less important, whereas the information expressed by the expression outside becomes relatively more important. We call such effects the **ascending and descending effects** of parentheses[10].

In a plain disjunction "A or B" where the two disjuncts are equally important, adding parentheses triggers the ascending and descending effects, making the disjunct outside the parentheses more significant than the one inside.

$$A \text{ or } B \overset{+(\,)}{\Longrightarrow} \overset{\uparrow}{A}(\text{or } \overset{\downarrow}{B})$$

This produces an **ordering** between A and B, which is a preference ordering if A is preferred to B (type I of disjunctions with parentheses, see Sect. 2.1), a plausibility or likelihood ordering if A is more likely than B (type II), and an appropriateness ordering over wording if A is more appropriate wording than B (type III).

Note that this account for the source of orderings also applies to other types of ordered disjunctions that trigger the ascending and descending effects.

The function of parentheses in "A (or B)" can also be understood by considering the underlying **information structure**. When both disjuncts A and B are focuses of the sentence, as determined by a QUD, parentheses act as a *non-focal marker*, rendering the item in parentheses non-focal.

$$[A]_F \text{ or } [B]_F \overset{+(\,)}{\Longrightarrow} [A]_F(\text{or } B)$$

Thus, the QUDs for "A or B" and "A (or B)" can differ, reflecting the difference in the focus of these disjunctions. For example, "Is this room for socializing?" can be a QUD for (1-a), but not for (2-a).

4 Reasoning Involving Ordered Disjunction

In this section, we study some patterns of reasoning involving ordered disjunction encoded by disjunctions with parentheses, focusing on the effects of the orderings between disjuncts conveyed by those disjunctions. In our examples from natural language, we use disjunctions with parentheses as representatives of ordered disjunctions, as the former are among the most natural constructions that encode the latter in written language.

4.1 Commutative Law

Classical boolean disjunction satisfies the commutative law: $A \vee B$ is semantically equivalent to $B \vee A$, meaning their semantic values are irrelevant to the order of disjuncts. However, as we mentioned in the introduction, the commutative law does not hold for an ordered disjunction "A (or B)."

[10] We believe that such effects can also be triggered by other means of expressing emphasis, for instance phrasal stress, intonation, underlining, boldface, italics, etc.

The semantic value of "A (or B)," according to our analysis, includes the ordering saying that A is preferred to B; the semantic value of "B (or A)," however, includes the converse ordering. Thus, "A (or B)" cannot be semantically equivalent to "B (or A)."

4.2 Pragmatic Inferences Triggered by Ordered Disjunction

We will show in this section that the orderings introduced by ordered disjunctions play a role in some pragmatic inferences from them, leading to new types of *free choice inferences* (FC, [18]) and *ignorance inferences* ([16]).

A **free choice inference** occurs when a conjunction of possibility modal statements is derived from a disjunctive modal statement. Consider the following examples:

(23) Deontic $\Diamond$-FC

a. You may go to the beach or to the cinema.

b. $\rightsquigarrow$ You may go to the beach and you may go to the cinema. ([18])

(24) Deontic $\Box$-FC

a. You ought to undergo surgery or take some medicine.

b. $\rightsquigarrow$ You are permitted to undergo surgery, and you are permitted to take some medicine.

(25) Epistemic $\Diamond$-FC

a. Mr. X might be in Victoria or in Brixton.

b. $\rightsquigarrow$ Mr. X might be in Victoria and he might be in Brixton. ([30])

(26) Epistemic $\Box$-FC

a. Baoyu must be from Beijing or Shanghai.

b. $\rightsquigarrow$ Baoyu might be from Beijing and Baoyu might be from Shanghai.

The statements in (a) in a modal context present multiple options to the audience, which are all viable choices. Similar derivation can also be made from plain disjunctive statements *without* modals. Such derivation is known as **ignorance inferences**, for example:

(27) Ignorance inferences

a. Baoyu is in Yihongyuan or Xiaoxiangguan.

b. $\rightsquigarrow$ Baoyu might be in Yihongyuan, and he might be in Xiaoxiangguan.

In classical modal logic, especially in standard deontic and epistemic frameworks, these inferences are logically invalid. Many theories suggest that these inferences stem from *pragmatic* effects rather than semantic entailment. For instance, FC and ignorance inferences are often treated as conversational implicatures, as discussed in studies by [22,24]; FC inferences have also been analyzed as scalar implicatures, see [10,15] among others. [2] introduces another perspective, proposing that such inferences are triggered by *neglect-zero* effects (see

Sect. 5.1). It suggests that these inferences arise from a tendency among speakers to construct representations of reality while systematically neglecting empty configurations.

Although these theories approach the inferences from different perspectives, they all acknowledge and assume that such derivations are not logical errors, but rather outcomes of pragmatic enrichment, summarized in the following principles:

- FC principle
  $\Box / \Diamond (A \vee B) \rightsquigarrow \Diamond A \wedge \Diamond B.$
- Ignorance principle
  $A \vee B \rightsquigarrow \Diamond A \wedge \Diamond B.$

When parentheses are added to disjunctive statements, giving rise to ordered interpretations, the derivation of FC and ignorance inferences from the resulting disjunctive statements *differs* from that in the case when there are no parentheses: a "middle" modal with a recommendation meaning can be derived for the disjunct outside the parentheses, while a possibility modal is derived for the disjunct inside the parentheses; this differs from classical FC and ignorance inferences for boolean disjunction. Consider the following examples:

(28) Deontic ordered $\Diamond$-FC

a. You may go to the beach (or to the cinema).

b. $\rightsquigarrow$ You are suggested to go to the beach, and you may go to the cinema.

(29) Deontic ordered $\Box$-FC

a. You ought to undergo surgery (or take some medicine).

b. $\rightsquigarrow$ You are strongly recommended to undergo surgery, but you can also take some medicine.

(30) Epistemic ordered $\Diamond$-FC

a. Mr. X might be in Victoria (or in Brixton).

b. $\rightsquigarrow$ It is more likely that Mr. X is in Victoria, but he might also be in Brixton.

(31) Epistemic ordered $\Box$-FC

a. Baoyu must be from Beijing (or Shanghai).

b. $\rightsquigarrow$ It is highly likely that Baoyu is from Beijing, but he might also be from Shanghai.

(32) Ordered ignorance inferences

a. Baoyu is in Yihongyuan (or Xiaoxiangguan).

b. $\rightsquigarrow$ It is highly likely that Baoyu is in Yihongyuan, but he might also be in Xiaoxiangguan.

In these examples, the orderings of options influence modal reasoning by adjusting the importance of each option, so these inferences derive from an ordered disjunction a basic possibility and a modality that combines possibility with importance. The latter will be denoted by the "middle" modality. We use $\boxminus$ to represent a middle modality that conveys a sense of strong suggestion or high likelihood. Similarly, we use $\diamondplus$ to represent a middle modality that conveys suggestion or likelihood. The intuition is that these middle modalities are semantically weaker than necessity modalities but stronger than possibility modalities. Furthermore, between the middle modalities, $\boxminus$ expresses a stronger attitude than $\diamondplus$.

For example, in (28), the middle modality "you are suggested/recommended" merges permission with preference encoded by the disjunction with parentheses. Similarly, in (30), "it is more likely" combines epistemic possibility with likelihood.

We argue that the middle modalities result from the ascending and descending effects triggered by parentheses, and other prosodic, typographical, or punctuational devices emphasizing some of the disjuncts. The ascending effects significantly influence the reasoning process. They not only elevate the importance of the information provided by A, the highlighted disjunct, but also change the attitude towards A, resulting in a modality stronger than basic possibility, i.e., a middle modality.

$$\Box / \Diamond\, (A \text{ or } B) \overset{+(\,)\&\mathtt{FC}}{\Longrightarrow} \overset{\uparrow\,\uparrow}{\Diamond A} \text{ and } \Diamond\, B$$

As observed above, the FC and ignorance principles for ordered disjunction vary from those for boolean disjunction. They are given as follows:

- Ordered FC principle
  - $\Box(A(\vee B)) \rightsquigarrow \boxminus A \wedge \Diamond B.$
  - $\Diamond(A(\vee B)) \rightsquigarrow \diamondplus A \wedge \Diamond B.$
- Ordered ignorance principle
  $A(\vee B) \rightsquigarrow \boxminus A \wedge \Diamond B.$

In addition, as highlighted by [5], FC effects disappear in negative contexts. We consider the example where Baoyu is explicitly prohibited from marrying either of two individuals:

(33) a. Baoyu is not allowed to marry Daiyu or Xiren.
b. ⇒ Baoyu is not allowed to marry Daiyu, and he is not allowed to marry Xiren.

Instead of implying a choice between prohibitions, (33-a) categorically states that Baoyu is forbidden from marrying both Daiyu and Xiren.

Similarly, the ordered FC triggered by ordered disjunction also disappears under negation, accompanying the disappearance of the ordering between disjuncts observed in Sect. 2.3:

(34) a. Baoyu is not allowed to marry Daiyu (or Xiren.)

b. ⇒ Baoyu is not allowed to marry Daiyu, and he is not allowed to marry Xiren.
c. # ⇝ Baoyu is not suggested to marry Daiyu, and he is not allowed to marry Xiren.

In this negative context, the middle modality of suggestion, typically derived in positive disjunctive statements, cannot be inferred.

5 Semantics for Ordered Disjunction

In this section, we propose a semantic framework specifically designed to capture ordered disjunction and its associated inferences discusses in the previous section. We will employ the Bilateral State-based Modal Logic (BSML, [2]) as our foundational framework, which analyzes FC as triggered by neglect-zero effects.

5.1 Neglect-Zero Effects

As discussed in Sect. 4.2, Aloni [2] introduces a cognitive tendency known as *neglect-zero*[11]. This tendency of favoring concrete models over abstract models significantly influences language comprehension and model representation by avoiding abstract configurations that validate sentences through *zero models*.

To illustrate this concept, consider the following models:

(35) There are some apples or a peach.
a. Verifier: {}
b. Falsifier: {}
c. Zero-models: 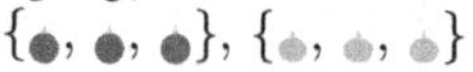

In the example, each set represents a model. (35-a) depicts a model that verifies the statement with a configuration of apples (red ones) and a peach (pink one), while (35-b) shows a model that falsifies the statement, as it contains neither apples nor peaches. (35-c) represents zero-models that verify the statement due to the presence of empty witnesses for one of the disjuncts. In these zero models, there are either only apples, or only peaches.

Aloni proposed the Bilateral State-based Modal Logic (BSML, [2]) to formalize the assumption of neglect-zero and the inferences triggered by it. BSML employs team semantics, which evaluates formulas at sets of possible worlds (such sets are known as information states/teams) rather than single worlds. BSML adopts bilateral semantics so there are two interpreting conditions for each formula, which are assertion and rejection, modeled by *support* (⊨) and *anti-support* (⫤), respectively:

[11] A number of arguments for the neglect-zero hypothesis can be found in [2,4]. In addition, this approach received further confirmation from a number of experimental studies, including a recent paper [12], which focuses on ignorance inferences.

- $\mathcal{M}, s \models \phi$ indicates that ϕ is assertable in state s, where $s \subseteq W$.
- $\mathcal{M}, s \dashv \phi$ indicates that ϕ is rejectable in state s, where $s \subseteq W$.

A key element in BSML is the non-emptiness atom (NE) from team logic. In team-based semantics, the empty state supports every *classical* formula, including contradictions. The interpretation of NE requires the supporting team to be non-empty, effectively excluding zero models from natural language interpretation.

The concept of neglect-zero can be formalized by a operation called *pragmatic enrichment*, defined recursively by the function $[\,]^+$ and the non-emptiness atom NE. For example, $[p]^+ = p \wedge \mathtt{NE}$, where p is a propositional letter.

Therefore in BSML, in addition to classical modal formulas ϕ, there are pragmatically enriched formulas $[\phi]^+$. It is important to note that free choice and ignorance inferences can be derived from these pragmatically enriched formulas, but not from an arbitrary formula.

- $[\Box(\phi \vee \psi)]^+ \models \Diamond\phi \wedge \Diamond\psi$
- $\Box(\phi \vee \psi) \not\models \Diamond\phi \wedge \Diamond\psi$
- $[\Diamond(\phi \vee \psi)]^+ \models \Diamond\phi \wedge \Diamond\psi$
- $\Diamond(\phi \vee \psi) \not\models \Diamond\phi \wedge \Diamond\psi$
- $[\phi \vee \psi]^+ \models \Diamond\phi \wedge \Diamond\psi$
- $\phi \vee \psi \not\models \Diamond\phi \wedge \Diamond\psi$

Further details on the logic and its applications in linguistics are discussed in [2,6].

5.2 BSML with the Ordered Disjunction

In this section, we define the ordered disjunction in the framework of BSML.

We define a specialized **language** $\mathcal{L}_O$ based on BSML language.

Definition 1 (Language $\mathcal{L}_O$). *Let $A = \{p, q, r, \ldots\}$ be a set of propositional letters. Formulas in $\mathcal{L}_O$ are formed as follows:*

$$\phi := p \mid \neg\phi \mid \phi \wedge \phi \mid \phi \vee \phi \mid \phi \mathbin{\vec{\vee}} \phi \mid \Box\phi \mid \boxminus\phi \mid \hat{\Diamond}\phi \mid \mathtt{NE}$$

where $p \in A$.

$\mathcal{L}_O$ extends BSML language by including both the ordered disjunction and the middle modality. As a result, there are two disjunction operators in $\mathcal{L}_O$. One is known as the *split disjunction* (or tensor disjunction) in team semantics. A split disjunction $\phi \vee \psi$ is supported by a state s if s can be partitioned into two substates, each supporting one of the disjuncts. The other one is the *ordered disjunction* $\vec{\vee}$, which further adds orderings to the split disjunction.

Additionally, $\mathcal{L}_O$ introduces the middle modality operators $\boxminus$ and $\hat{\Diamond}$, which as discussed in Sect. 4.2 are products of pragmatic inferences involving ordered disjunctions.

Pragmatic enrichment can also be defined for these new operators.

Definition 2 (Pragmatic Enrichment of $\mathcal{L}_O$). *The pragmatic enrichment function is a mapping* $[\ ]^+$ *from the* NE*-free fragment of* $\mathcal{L}_O$ *to* $\mathcal{L}_O$ *such that:*

- $[\phi \veebar \psi]^+ = ([\phi]^+ \veebar [\psi]^+) \wedge \texttt{NE}$
- $[\boxminus\phi]^+ = \boxminus[\phi]^+ \wedge \texttt{NE}$
- $[\Diamond\phi]^+ = \Diamond[\phi]^+ \wedge \texttt{NE}$
- *Clauses for other operators are the same as corresponding clauses for the pragmatic enrichment of BSML language.*

A **model** for BSML is a Kripke-style model which is a triple $\mathcal{M} = \langle W, R, V\rangle$, where

1. W is a set of possible worlds,
2. R is an accessibility relation on W,
3. $V : A \times W \to \{0, 1\}$ is a world-dependent valuation function.

BSML formulas are interpreted in models $\mathcal{M} = \langle W, R, V\rangle$ with respect to information states $s \subseteq W$.

We now extend the model to include an ordering, commonly employed in deontic logic and preference logic and known as the betterness relation (see [7,8] among others). This would enable our semantics to capture types I and II of disjunctions with parentheses, which, according our discussion in Sect. 2.2, both introduce a context-dependent ordering to the semantics.

Definition 3 (Ordered BSML Models). *An ordered BSML model is a quadruple* $\mathcal{M} = \langle W, R, \preceq, V\rangle$*, where*

1. $\langle W, R, V\rangle$ *is a BSML model,*
2. $\preceq \subseteq W \times W$ *is a reflexive and transitive relation ("betterness" pre-order) over possible worlds.*

For any possible worlds w_1 and w_2, $w_1 \preceq w_2$ can be read as "w_2 is at least as good as w_1."

We further define the set of successors of a possible world w under R and the best part of an information state s according to $\preceq$:

- $R[w] = \{v \in W \mid Rwv\}$
- $\texttt{BEST}(s) = \{w \in s \mid \text{for all } w' \in s,\ w \preceq w' \text{ implies } w' = w\}$

In fact, for every information state s, $\texttt{BEST}(s)$ is a $\preceq$-antichain because it is a set of $\preceq$-maximal elements which are pairwise incomparable. Moreover, each $\preceq$-antichain equals to $\texttt{BEST}(s)$ for some information state s.

We then give **semantic clauses** for $\mathcal{L}_O$ formulas.

Definition 4 (Semantic Clauses). *The support* ($\models$) *and anti-support* ($=\!|$) *relations of an* $\mathcal{L}_O$ *formula* ϕ *by an information state* s *in an ordered BSML model* $\mathcal{M} = \langle W, R, \preceq, V\rangle$ *is defined as follows:*

$\mathcal{M}, s \vDash p$	iff for all $w \in s, V(w,p) = 1$
$\mathcal{M}, s \dashv p$	iff for all $w \in s, V(w,p) = 0$
$\mathcal{M}, s \vDash \neg\phi$	iff $\mathcal{M}, s \dashv \phi$
$\mathcal{M}, s \dashv \neg\phi$	iff $\mathcal{M}, s \vDash \phi$
$\mathcal{M}, s \models \phi \wedge \psi$	iff $\mathcal{M}, s \models \phi$ and $\mathcal{M}, s \models \psi$
$\mathcal{M}, s \dashv \phi \wedge \psi$	iff there exist $t, t' \subseteq s$ s.t. $t \cup t' = s$, $\mathcal{M}, t \dashv \phi$, and $\mathcal{M}, t' \dashv \psi$
$\mathcal{M}, s \vDash \phi \vee \psi$	iff there exist $t, t' \subseteq s$ s.t. $t \cup t' = s$, $\mathcal{M}, t \vDash \phi$, and $\mathcal{M}, t' \vDash \psi$
$\mathcal{M}, s \dashv \phi \vee \psi$	iff $\mathcal{M}, s \dashv \phi$ and $\mathcal{M}, s \dashv \psi$
$\mathcal{M}, s \vDash \phi \vee\!\!\vee \psi$	iff there exists $t \subseteq s$ s.t. $\texttt{BEST}(s) \cup t = s$, $\mathcal{M}, \texttt{BEST}(s) \vDash \phi$, and $\mathcal{M}, t \vDash \psi$
$\mathcal{M}, s \dashv \phi \vee\!\!\vee \psi$	iff $\mathcal{M}, s \dashv \phi$ and $\mathcal{M}, s \dashv \psi$
$\mathcal{M}, s \vDash \Box\phi$	iff for all $w \in s$, $\mathcal{M}, R[w] \vDash \phi$
$\mathcal{M}, s \dashv \Box\phi$	iff for all $w \in s$, there exists $t \subseteq R[w]$ s.t. $t \neq \emptyset$ and $\mathcal{M}, t \dashv \phi$
$\mathcal{M}, s \vDash \boxminus\phi$	iff for all $w \in s$, $\mathcal{M}, \texttt{BEST}(R[w]) \models \phi$
$\mathcal{M}, s \dashv \boxminus\phi$	iff for all $w \in s$, there exists $t \subseteq \texttt{BEST}(R[w])$ s.t. $t \neq \emptyset$ and $\mathcal{M}, t \dashv \phi$
$\mathcal{M}, s \vDash \diamondplus\phi$	iff for all $w \in s$, there exists $t \subseteq R[w]$ s.t. $t \neq \emptyset$ and $\mathcal{M}$, $\texttt{BEST}(t) \models \phi$
$\mathcal{M}, s \dashv \diamondplus\phi$	iff for all $w \in s$ and all non-empty $t \subseteq R[w]$, $\mathcal{M}, \texttt{BEST}(t) \dashv \phi$[1]

The possibility modal $\Diamond$ in BSML is defined as the dual of $\Box$:
$\mathcal{M}, s \vDash \Diamond\phi$ iff for all $w \in s$, there exists $t \subseteq R[w]$ s.t. $t \neq \emptyset$ and $\mathcal{M}, t \vDash \phi$
$\mathcal{M}, s \dashv \Diamond\phi$ iff for all $w \in s$, $\mathcal{M}, R[w] \dashv \phi$

We explain the intuitive meaning of the middle modalities in epistemic contexts, where $\boxminus\phi$ indicates that ϕ is highly likely. Given an information state s, $s \models \boxminus\phi$ means that for every world w in s, ϕ is supported by the best part of the set of *all* possible worlds the agent considers possible if w were the actual world. On the other hand, $\diamondplus\phi$ signifies that ϕ is more likely, which is semantically weaker than $\boxminus\phi$. Specifically, $s \models \diamondplus\phi$ indicates that in every world w in s, ϕ is supported by the best part of a set of *some* possible worlds that the agent considers possible if w were the actual world. The intuitive meaning of $\boxminus$ and $\diamondplus$ in deontic contexts is similar.

The following fact about $\mathcal{L}_O$ **modals** can be proven straightforwardly.

Fact 1.

- $\boxminus\phi \vDash \Diamond\phi$, *assuming seriality of* R
- $\diamondplus\phi \vDash \Diamond\phi$

To formulate the relationship between $\Box$ and $\boxminus$, we need to define downward closure, a key property in team-based semantics (see [3,27]).

Definition 5 (Downward Closure). *An $\mathcal{L}_O$ formula ϕ is downward closed, iff $\mathcal{M}, s \models \phi$ and $t \subseteq s$ imply $\mathcal{M}, t \models \phi$.*

It is easy to show the following fact.

[12] This is equivalent to: for any $w \in s$ and any non-empty $\preceq$-antichain $t \subseteq R[w]$, $\mathcal{M}, t \dashv \phi$.

Fact 2. *Assuming downward closure of ϕ, $\Box\phi \vDash \boxminus\phi$.*

As for the two types of **disjunction**, the split disjunction requires that the information state s can be split into two subsets, each supporting one disjunct. The ordered disjunction further introduces an ordering constraint, stipulating that the subset supporting the first disjunct must be $\texttt{BEST}(s)$. In other words, the ordered disjunction splits the information state according to the ordering. As expected, the ordered disjunction is strictly stronger than the split disjunction and blocks the commutative law, as shown by the following fact, whose proof is straightforward.

Fact 3.

- $\phi \veebar \psi \vDash \phi \vee \psi$
- $\phi \vee \psi \nvDash \phi \veebar \psi$
- $\phi \veebar \psi \nvDash \psi \veebar \phi$

In addition, the following facts about **downward closure** can be established.

Fact 4. *$\boxminus\phi$ is downward closed.*

Proof. The proof follows from the semantics of $\boxminus$. Specifically, since $\mathcal{M}, s \models \boxminus\phi$ implies that for all $w \in s$, $\texttt{BEST}(R[w])$ supports ϕ, it follows that for any $t \subseteq s$, and for all $w \in t$, $\texttt{BEST}(R[w])$ also supports ϕ.

□

Fact 5. *$\Diamond\phi$ is downward closed.*

Proof. The proof is analogous to that of the downward closure of $\boxminus\phi$, and is therefore omitted here.

□

In team semantics and BSML, split disjunctions (a.k.a. tensor disjunctions), represented by $\phi \vee \psi$, are downward closed, assuming the downward closure of each disjunct (see [3,27]). The ordered disjunction proposed in this paper is a type of disjunction based on the split disjunction but restricts the method of splitting: one split must be the best part of the current information state. As $s' \subseteq s$ does not generally imply $\texttt{BEST}(s') \subseteq \texttt{BEST}(s)$, this restriction results in $\phi \veebar \psi$ not being downward closed, even if both ϕ and ψ are downward closed.

Fact 6. *There exist downward closed formulas ϕ and ψ such that $\phi \veebar \psi$ is not downward closed.*

Proof. A counterexample can be constructed as follows: Let $\phi = p$, $\psi = \top$, and $\mathcal{M} = \langle W, R, \preceq, V\rangle$ be an ordered BSML model, where

1. $W = \{w_1, w_2, w_3\}$,
2. $\preceq = \{\langle w_2, w_1\rangle, \langle w_3, w_2\rangle, \langle w_3, w_1\rangle, \langle w_1, w_1\rangle, \langle w_2, w_2\rangle, \langle w_3, w_3\rangle\}$,
3. $V(p, w_1) = 1$, $V(p, w_2) = V(p, w_3) = 0$.

Let $s = W$. Since $\texttt{BEST}(s) = \{w_1\}$, $\mathcal{M}, \{w_1\} \models p$, and $\mathcal{M}, \{w_2, w_3\} \models \top$, we have $s \models p \mathbin{\ovee} \top$.

Let $s' = \{w_2, w_3\} \subseteq s$. Note that $\texttt{BEST}(s') = \{w_2\}$, but $\{w_2\} \not\models p$, so $s' \not\models p \mathbin{\ovee} \top$.

□

5.3 Pragmatic Inferences Triggered by the Ordered Disjunction

The semantics of the ordered disjunction in our framework correctly predicts the pragmatic inferences involving ordered disjunction that we discussed in Sect. 4.2.

In BSML, free choice inferences and ignorance inferences can be derived from pragmatically enriched formulas, but not from any split disjunction. This also applies to the ordered disjunction.

Fact 7 (Ordered *FC*).

1. $[\Box(\phi \mathbin{\ovee} \psi)]^+ \models \boxminus\phi \wedge \Diamond\psi$
2. $\Box(\phi \mathbin{\ovee} \psi) \not\models \boxminus\phi \wedge \Diamond\psi$
3. $[\Diamond(\phi \mathbin{\ovee} \psi)]^+ \models \diamondminus\phi \wedge \Diamond\psi$
4. $\Diamond(\phi \mathbin{\ovee} \psi) \not\models \diamondminus\phi \wedge \Diamond\psi$

Proof.

1. Suppose $\mathcal{M}, s \models [\Box(\phi \mathbin{\ovee} \psi)]^+$, which implies that $\mathcal{M}, s \models \Box(([\phi]^+ \mathbin{\ovee} [\psi]^+) \wedge \texttt{NE})$. It follows that for all $w \in s$, $R(w) \models [\phi]^+ \mathbin{\ovee} [\psi]^+$. So there exists $t \subseteq R(w)$ such that $\texttt{BEST}(R[w]) \cup t = R[w]$, $\texttt{BEST}(R(w)) \models [\phi]^+$, and $t \models [\psi]^+$. Then $\texttt{BEST}(R(w)) \models \phi, t \models \psi$, and $t \neq \emptyset$. Therefore, $s \models \boxminus\phi$ and $s \models \Diamond\psi$.
2. Let $\mathcal{M}$ be any ordered BSML model. Trivially, $\mathcal{M}, \emptyset \models \Box(\varphi \mathbin{\ovee} \psi)$. However, $\emptyset \not\models \Diamond\psi$, so $\emptyset \not\models \boxminus\phi \wedge \Diamond\psi$.
3. Suppose $\mathcal{M}, s \models [\Diamond(\phi \mathbin{\ovee} \psi)]^+$, which implies that $\mathcal{M}, s \models \Diamond(([\phi]^+ \mathbin{\ovee} [\psi]^+) \wedge \texttt{NE})$. It follows that for all $w \in s$ there exists a non-empty subset $t \subseteq R[w]$ such that $t \models [\phi]^+ \mathbin{\ovee} [\psi]^+$. It then follows that there exists $t' \subseteq t$ such that $\texttt{BEST}(t) \cup t' = t$, $\texttt{BEST}(t) \models \phi, \texttt{BEST}(t) \neq \emptyset, t' \models \psi$, and $t' \neq \emptyset$. Based on the semantics of $\diamondminus$, it follows that $s \models \diamondminus\phi$. Since $t' \subseteq t \subseteq R[w]$, we have $s \models \Diamond\psi$.
4. Let $\mathcal{M} = \langle W, R, \preceq, V \rangle$ be an ordered BSML model, where
   - $W = \{w\}$,
   - $R = \{\langle w, w \rangle\}$,
   - $\preceq = \{\langle w, w \rangle\}$,
   - $V(w, p) = 1$.

   Then $\mathcal{M}, \{w\} \models \Diamond(p \mathbin{\ovee} \neg p)$, but $\{w\} \not\models \Diamond\neg p$.

□

Ordered ignorance inferences intuitively bring out the intrinsic epistemic meaning of the current information state, so they require the information state to be an epistemic state and satisfy some epistemic closure conditions, formulated in BSML by Aloni ([2]) as state-basedness.

Definition 6 (State-basedness). *Let* $\mathcal{M} = \langle W, R, \prec, V \rangle$ *be an ordered BSML model and* $s \subseteq W$. *R is state-based in* $\langle M, s \rangle$ *iff for all* $w \in s$, $R[w] = s$.

Fact 8 (Ordered Ignorance Inferences). *Assuming R is state-based,*

1. $[(\phi \vee\!\!\!\vee \psi)]^+ \vDash \boxminus\phi \wedge \Diamond\psi$
2. $(\phi \vee\!\!\!\vee \psi) \nvDash \boxminus\phi \wedge \Diamond\psi$

Proof

1. Suppose $\mathcal{M}, s \vDash [(\phi \vee\!\!\!\vee \psi)]^+$, from which follows that $\mathcal{M}, s \vDash ([\phi]^+ \vee\!\!\!\vee [\psi]^+) \wedge \mathtt{NE}$. Then there exists $t \subseteq s$ such that $\mathtt{BEST}(s) \cup t = s$, $\mathtt{BEST}(s) \vDash \phi, t \vDash \psi$, and $t \neq \emptyset$. Because of the state-basedness of R, we conclude that $s \vDash \boxminus\phi$ and $s \vDash \Diamond\psi$.
2. Take $\mathcal{M}$ in the proof of Fact 7. Note that $\mathcal{M}, \{w\} \models p \vee\!\!\!\vee \neg p$, but $\{w\} \nvDash \Diamond\neg p$.

□

The following fact shows that ordered FC and ordered ignorance inferences of the ordered disjunction disappear under negation.

Fact 9 (Dual Prohibition).

- $[\neg \Diamond (\phi \vee\!\!\!\vee \psi)]^+ \vDash \neg \Diamond \phi \wedge \neg \Diamond \psi$
- $[\neg(\phi \vee\!\!\!\vee \psi)]^+ \vDash \neg \Diamond \phi \wedge \neg \Diamond \psi$, *assuming R is state-based.*

We skip the proof of this fact, which is straightforward.

It is important to note that our current proposal for the semantics of ordered disjunction does not fully capture disjunctions with parentheses of type III, which involve orderings between co-referential signifiers, and thus require an extension of our framework to distinguish between referents and signifiers. One possible approach is to extend our logic to a two-sorted modal logic whose two sorts are referents and signifiers. Another possible approach is to adopt the framework of conceptual cover discussed in [1]. We leave this to future work.

6 Conclusion

In this paper, we explored ordered interpretations of disjunctive expressions in natural language, specifically focusing on disjunctions with parentheses. Our exploration highlighted that while the standard boolean reading applies to basic disjunction, the addition of parentheses, similar to prosodic devices (e.g., phrasal stress) and typographical means (e.g., boldface), can introduce context-dependent ordered interpretations. We categorized these interpretations into three distinct types: preference, likelihood, and appropriateness of wording. By analyzing various examples, we demonstrated how the pragmatic effects of parentheses, namely the ascending and descending effects, contribute to these ordered readings.

Furthermore, we proposed a semantics to capture these interpretations, integrating ordered disjunction into Bilateral State-based Modal Logic (BSML). Our framework characterizes logical characteristics of ordered disjunction and supports the derivation of free choice and ignorance inferences.

Our proposed semantics can capture the first two types of disjunctions with parentheses – preference and likelihood – but not the third type, the formalization of which is reserved for future work.

Our findings suggest that disjunctions in natural language can encode more information than truth of either of the disjuncts. Future research directions include expanding on our model to incorporate additional pragmatic factors and exploring the interplay between disjunction and other linguistic elements.

Acknowledgments. We would like to express our sincere gratitude to the audience of the Fourth Tsinghua Interdisciplinary Workshop on Logic, Language and Meaning (TLLM 2024) for their valuable feedback and discussions, as well as to the two anonymous reviewers of the TLLM proceedings for their insightful comments and suggestions. Our heartfelt thanks go to Professor Fenrong Liu and Professor Maria Aloni for their advice, with special appreciation for Professor Maria Aloni's detailed guidance. We are also grateful to Yumin Ji for the discussions and for her presentation at the TLLM conference. Finally, we would like to thank Søren Brinck Knudstorp for his helpful discussions on the formalization aspect of this paper.

References

1. Aloni, M.: Individual concepts in modal predicate logic. J. Philos. Log. **34**, 1–64 (2005)
2. Aloni, M.: Logic and conversation: the case of free choice. Semant. Pragmat. **15**, 5 (2022)
3. Aloni, M., Anttila, A., Yang, F.: State-based modal logics for free choice (2024). https://arxiv.org/abs/2305.11777
4. Aloni, M., van Ormondt, P.: Modified numerals and split disjunction: the first-order case. J. Logic Lang. Inform. **32**, 539–567 (2023)
5. Alonso-Ovalle, L.: Disjunction in Alternative Semantics. Ph.D. thesis, University of Massachusetts, Amherst (2006)
6. Anttila, A.: The Logic of Free Choice: Axiomatizations of State-based Modal Logics. Master's thesis, ILLC, University of Amsterdam (2021)
7. van Benthem, J., Grossi, D., Liu, F.: Deontics = betterness + priority. In: Governatori, G., Sartor, G. (eds.) DEON 2010. LNCS (LNAI), vol. 6181, pp. 50–65. Springer, Heidelberg (2010). https://doi.org/10.1007/978-3-642-14183-6_6
8. van Benthem, J., Liu, F.: Dynamic logic of preference upgrade. J. Appl. Non-Classical Logics **17**(2), 157–182 (2007)
9. Brewka, G.: Logic programming with ordered disjunction. In: Dechter, R., Kearns, M., Sutton, R. (eds.) Proceedings of AAAI-2002, pp. 100–105. American Association for Artificial Intelligence, Menlo Park, CA (2002)
10. Chierchia, G., Fox, D., Spector, B.: The grammatical view of scalar implicatures and the relationship between semantics and pragmatics. In: Maienborn, C., von Heusinger, K., Portner, P. (eds.) Semantics: An International Handbook of Natural Language Meaning. de Gruyter (2011)
11. Davies, M.: The corpus of contemporary American English (COCA). https://www.english-corpora.org/coca/ (2008)

12. Degano, M., et al.: The ups and downs of ignorance (2023). manuscript
13. Fauconnier, G.: Pragmatic scales and logical structure. Linguist. Inquiry **6**(3), 353–375 (1975)
14. Fløttum, K.: The self and the others: polyphonic visibility in research articles. Int. J. Appl. Linguist. **15**(1), 29–44 (2005)
15. Fox, D.: Free choice and the theory of scalar implicatures. In: Sauerland, U., Stateva, P. (eds.) Presupposition and Implicature in Compositional Semantics, pp. 71–120. Palgrave MacMillan, Hampshire (2007)
16. Grice, H.P.: Studies in the Way of Words. Harvard University Press, Cambridge, MA (1989)
17. Hyland, K.: Metadiscourse: what is it and where is it going? J. Pragmat. **113**, 16–29 (2017)
18. Kamp, H.: Free choice permission. Proc. Aristot. Soc. **74**, 57–74 (1973)
19. Maekawa, K., et al.: Balanced corpus of contemporary written Japanese. Lang. Resour. Eval. **48**, 345–371 (2014)
20. Potts, C.: The Logic of Conventional Implicatures. Oxford University Press, Oxford (2005)
21. Roberts, C.: Information structure: Towards an integrated theory of formal pragmatics. In: Yoon, J.H., Kathol, A. (eds.) Papers in Semantics, OSU Working Papers in Linguistics, vol. 49, pp. 91–136. The Ohio State University, Columbus, OH (1996)
22. Sauerland, U.: Scalar implicatures in complex sentences. Linguist. Philos. **27**, 367–391 (2004)
23. Schneider, S.: Parenthesis: Fundamental features, meanings, discourse functions and ellipsis. In: Kluck, M., Ott, D., de Vries, M. (eds.) Parenthesis and Ellipsis: Cross-Linguistic and Theoretical Perspectives, pp. 277–300. De Gruyter Mouton, Berlin, München, Boston (2015)
24. Simons, M.: A Gricean view on intrusive implicatures. In: Petrus, K. (ed.) Meaning and Analysis: New Essays on H. Paul Grice, pp. 138–169. Palgrave (2010)
25. Stevenson, A.: Oxford Dictionary of English. Oxford University Press, Oxford, third edn (2010)
26. Xiang, M., Kramer, A., Nordmeyer, A.E.: An informativity-based account of negation complexity. J. Exp. Psychol. Learn. Mem. Cogn. **46**(10), 1857–1867 (2020)
27. Yang, F., Väänänen, J.: Propositional team logics. Ann. Pure Appl. Logic **168**(7), 1406–1441 (2017)
28. Zhan, W., Guo, R., Chang, B., Chen, Y., Chen, L.: The building of the CCL corpus: its design and implementation. Corpus Linguist. **6**(1), 71–86 (2019)
29. Zhan, W., Guo, R., Chen, Y.: The CCL corpus of chinese texts. Center for Chinese Linguistics (CCL) of Peking University. http://ccl.pku.edu.cn:8080/ccl_corpus (2003)
30. Zimmermann, E.: Free choice disjunction and epistemic possibility. Nat. Lang. Seman. **8**, 255–290 (2000)

Negation, Disjunction, and Choice Questions

Reflections on Yuen Ren Chao on Chinese Logical Expressions

Byeong-uk Yi(✉)

University of Toronto, Toronto, Canada
b.yi@utoronto.ca
https://philosophy.utoronto.ca/directory/byeong-uk-yi/

Abstract. In an article published in 1955, "Notes on Chinese grammar and logic", Yuen Ren Chao gives one of the earliest discussions of logical expressions of Chinese. The article has many interesting and insightful remarks on expressions of logical notions in natural languages. This article has reflections on two of them. One concerns natural language expressions for *negation*; the other concerns expressions for *disjunction* and *choice questions* in various languages. In discussing the remarks, this article argues that Korean has a *sentential negation device*, which derives from a negative copula, and develops Chao's analysis of Chinese choice interrogatives (the *juxtaposition analysis*) to extend it to English, Japanese, and Korean choice interrogatives.

Keywords: Yuen Ren Chao · logical expression · Chinese · Korean · Japanese · negation · copula · disjunction · interrogative disjunction · the adverbial 'or' · alternative question · choice interrogative · the juxtaposition analysis · asyndetic · syndetic · coordinator · coordination

Yuen Ren Chao (1892–1982), an influential linguist who taught at Tsinghua University in early periods of his academic career (1920–1921, 1925–1938),[1] published an article titled "Notes on Chinese grammar and logic" in *Philosophy East*

[1] See, e.g., LaPolla (2017) for his biography.

The first part of this article stems from a short presentation on Chao's 1955 article in *The 4th Tsinghua Interdisciplinary Workshop on Logic, Language, and Meaning* (TLLM 2024). I wish to thank the audience, including David Huang, for useful comments and discussions. I would also like to thank Chungsik Bak for help on drafts for this article Bernard Katz and, Youich Matsusaka for discussions on English and Japanese choice interrogatives, and two anonymous referees for helpful comments on an earlier version. Special thanks are due to the organizing committee for the proceedings of TLLM 2024 for allowing me to write this article for the proceedings, and for their patience, and to Jialiang Yan and Joe Hung for discussions on delicate details of Chinese interrogatives. Needless yet important to say, I am solely responsible for any remaining errors and infelicities.

J. Yan et al. (Eds.): TLLM 2024, LNCS 15410, pp. 149–187, 2025.
https://doi.org/10.1007/978-3-031-86054-6_8

and West in 1955.[2] In the article, he discusses "the ways in which some elementary logical notions find expression in the Chinese language" (Chao 1955, 31). The discussions, where he refers to Russell and Whitehead's *Principia Mathematica* (1925) and Lewis and Langford's *Symbolic Logic* (1932),[3] are among the earliest, and *the* earliest I know of, of discussions of logical expressions of Chinese and other East Asian languages informed by the development of modern logic. They include interesting and insightful remarks that compare and contrast English and Chinese logical expressions. In this article, I reflect on two of them. One concerns natural language expressions for *negation*, the other expressions for *disjunction* and *choice questions* in various languages. While discussing the remarks, I argue that Korean has a *sentential negation device*, which derives from a negative copula (§1). And I develop Chao's analysis of Chinese choice interrogatives (the *juxtaposition analysis*) to extend it to English, Japanese, and Korean choice interrogatives (§2).[4]

1 Negation in Symbolic and Natural Languages

Symbols for negation used in modern logic (e.g., $-$, $\sim$, $\neg$, N, overbar) are *sentential operators*. They apply to entire sentences to yield their contradictories. Applying, e.g., $\sim$ to p, $(p \vee q)$, and $\exists x(Px \wedge Qx)$ yields the negations $\sim p$, $\sim (p \vee q)$, and $\sim \exists x(Px \wedge Qx)$, respectively, and the latter are true (or false) if and only if the former are false (or true). In natural languages, by contrast, the main expressions for negations (or negatives), e.g., the English 'not', usually apply to verbs (or predicates). Applying the negative particle 'not' to 'Socrates is snub-nosed', for example, yields 'Socrates is *not* snub-nosed', where the particle modifies the copula 'is'.[5] It is usual to take the latter to be the negation of the former and thus its contradictory, but adding the negative particle to the verbs (or predicates) does not always yield contradictories. For example, 'Some Greeks are snub-nosed' and 'Some Greeks are *not* snub-nosed' are not contradictories (they can both be true), and the latter is not considered the *negation* of the former.[6] If so, are there any natural language devices that, like negation symbols of symbolic languages, apply to entire sentences to form their negations (and thus contradictories)? Chao (1955, 32) raises this question in discussing Chinese expressions for negation.

---

[2] The article was "read before the XXIIIrd Congress of Orientalists, Cambridge, England, August 23, 1954" (Chao 1955, 31) before publication in 1955.

[3] He also refers to Fuh Yan's Chinese translation of Mill's *System of Logic* (Yan 1923).

[4] The discussion of *disjunction* and *choice interrogatives* in §2 does not depend on that of *negation* in §1, which readers interested in choice interrogatives might skip.

[5] See Aristotle's *Categories* (1963, 13b27–33). For a historical survey, see, e.g., Horn (1989, Chapter 1).

[6] Using 'no', another negative particle in English, we can get a contradictory of the former: '*No* Greeks are snub-nosed.' And we can get the negation of 'Every Greek is snub-nosed' by adding 'not' in front of 'every' or the entire sentence, as in '*Not* every Greek is snub-nosed'.

1.1 Sentential Negation in English and Chinese[7]

On Chinese negations, Chao says:

> A statement is denied by putting the general negative adverb *bu* ... before the predicate. Thus, if a proposition takes the form *S P* ... its contradictory ... will read: *S bu P*. For example, ... *Ta bu chy*, "He does not eat." (Chao 1955, 32)[8]

In this respect, Chinese is like English except that the English 'not' can be taken to be added *after* verbs (including auxiliary verbs, e.g., 'do') while the Chinese *bù* is added *before* them.[9] And he points out that the Chinese *bu*, like the English 'not', differs from negation symbols of symbolic languages: in "the usual logistic notation", by contrast, "the sign for the negation" is placed "before the whole sentence", which is "conducive to systematic simplicity" (Chao 1955, 32). And he adds that "neither Chinese nor any other natural language I know of does this systematically" (Chao 1955, 32).

With this remark, he seems to suggest a negative answer to the question whether any natural language has a device that, like negation symbols of symbolic languages, can systematically turn declarative sentences to their *negations* and thus *contradictories*. To raise this question, he assumes an incidental feature of negation symbols of usual symbolic languages: they are *sentence-frontal*, i.e., added before the negated sentences to figure at the beginning of their negations. But an important issue is not whether the negation symbol is placed *before* negated sentences but whether it directly applies to *the entire sentences* to form their negations or applies to some *special components* thereof (e.g., verbs or predicates, quantifiers, adverbs). While it is usual to add negation symbols (e.g., $\sim$) in front of the sentences they govern in usual symbolic languages,[10] it is straightforward to modify them to devise symbolic languages where negation symbols are placed *after* the negated sentences, as in $p \sim$, $(p \vee q) \sim$, $(p \wedge q) \sim\sim$, etc.[11]

---

[7] The discussions of Chinese in §1.1, §2.1, and §2.5 rely heavily on discussions on Chinese with Joe Hung, whom I wish to thank for patient instructions on delicate details of Chinese. Still, I am solely responsible for any error in the discussions, which includes my rephrasing and interpretation of his examples and explanations.

[8] And he points out that "There is no adjective in Chinese corresponding to 'no' " (Chao 1955, 32). In this respect, Chinese is the same as a variety of languages, including Korean and Arabic.

[9] See, e.g., Jespersen (1933, 297ff). McCawley says that the negative particle 'not', which "generally expresses negation", is usually added (essentially) to the verb so that "for example, *Cincinnati isn't in Mongolia* expresses the negation of the proposition expressed by *Cincinnati is in Mongolia*" (1981, 16). I add 'essentially' because in "John doesn't love his wife" (McCawley 1981, 16), for example, 'not' is not directly added to 'love' (or 'love his wife'). In subsequent discussions, however, I usually omit the qualification.

[10] In some symbolic languages, the negation symbol is added as the bar over the sentences they govern, as in $\overline{p}$ and $\overline{p \wedge \overline{q}}$.

[11] In reverse Polish notation, operator symbols, including the negation symbol, is added after the operands.

The negation symbols in such languages would be similar to negation symbols of usual symbolic languages and differ from the English 'not' and the Chinese *bù*, in being *sentential negation devices*, i.e., symbols or expressions forming negations by directly applying to entire sentences, including sentential compounds (e.g., $(p \vee q)$, $(p \wedge q)$). Similarly, some natural languages might have sentential negation devices added at (or near) the end of the negated sentences. So I think the interesting question Chao meant to raise is whether or not there are natural languages with *sentential negation* devices *regardless of where* they are added. I think the answer to this question is yes.

In works in modern logic, it is usual to explain that negation symbols of symbolic languages amount to the English 'it is not the case that', for 'It is not the case that Sam likes coffee', for example, is a contradictory of 'Sam likes coffee.'[12] But this does not mean that 'it is not the case that' is a *phrase* for forming sentential negations, for the construction is not a *phrase* or *syntactic constituent* of, e.g., 'It is not the case that Sam likes coffee', which does not result from combining the putative counterpart of $\sim$ with 'Sam likes coffee'—this sentence combines with 'that' to yield the *that*-clause 'that Sam likes coffee', which is a constituent of the entire sentence (the extraposed subject matching the anticipatory 'it'),[13] where the negative device is the particle 'not' that is added to 'is'. One might hold that essentially the same holds for 'not that' if it is used merely as a short for 'it is not the case that' to yield contradictories of subsequent sentences.[14]

Chinese has cousins of the two English constructions:

(a) *Shìshí bìngfēi* ... 'It is not a fact (that) ... (literally, the fact is not (that) ...)'.
(b) *bùshì* 'not correct, not that'.[15]

One might hold that it is problematic to consider *shìshí bìngfēi* a sentential negation device essentially for the same reason that it is controversial to take its English cousin to be one: *bìngfēi* 'be (actually) not' is a *negative copula* and does not directly combine with *shìshí* 'fact' in the '*Shìshí bìngfēi* . . .' construction, where the copula is to be followed by a complement (e.g., *rúcǐ* 'like this, so'), as

[12] See, e.g., Quine (1982, 9), Strawson (1952, 79), and Kalish, Montague, and Mar (1980, 2).

[13] In generative grammar, 'that' is considered the *complementizer* and the 'that'-clause a *complementizer clause*, which can figure as the subject of a sentence.

[14] Note that 'not that' is usually used as an idiom for suggesting that the speaker does not mean to suggest the truth of the subsequent sentence (as when one says, 'She wouldn't tell me how much it cost—not that I was really interested'), not as a construction for forming a straightforward contradictory of the subsequent sentence. (The example in parenthesis is from the online Cambridge Dictionary: https://dictionary.cambridge.org/dictionary/english/not-that).

[15] The former involves *shìshí* 'fact' and *bìngfēi* 'be (actually) not', and the latter *bù* 'not' and *shì* 'correct'. Thanks are due to David Huang for drawing my attention to the former construction, which Chao (1955) does not mention.

in *Shìshí bìngfēi rúcǐ* 'It is not so (literally, the fact is not like this)',[16] where *shìshí* and *bìngfēi* do not form a phrase any more than the English 'the fact' and 'is not' do in 'The fact is not like this.' While suggesting that natural languages do not have a systematic device for sentential negation, Chao takes note of *bùshì*:

> The nearest to that [viz., a sentential negation symbol] in Chinese is the . . . predicate *Bush* . . . (*sh*) "Not that . . . (but)," which is the form used for contrast or for forestalling an assertion to the contrary. (Chao 1955, 32)

Here he seems to take *bùshì* to fall short of qualifying as a sentential negation device because it is used in limited contexts (e.g., those in which a contrast or failure of expectation is suggested).

English and Chinese, we have seen, have systematic ways to turn declarative sentences to their negations or contradictories: we can add 'It is not a fact that . . .' (English) or *shìshí bìngfēi* ... (Chinese). They do so essentially by yielding statements to the effect that the sentences in question are not true or do not state a fact, and it is not clear that they can be considered sentential negation devices. It is hard to take them to form constituents of the resulting sentences or designed specifically for forming negations. But I think some natural languages (e.g., Korean, Japanese) have ways of forming negations that can more plausibly be considered sentential negation devices.

1.2 Sentential Negation in Korean

Korean has several negation devices.[17] It is useful for the present purpose to compare two of them:

(a) The adverb *an* 'not'.
(b) The verb *ani-ta* 'be not'.[18]

---

[16] The example is from the entry on *bìngfēi* in Kleeman and Yu (2010, Chinese-English Dictionary 49), where the translation given is "That is not actually the case." The copula phrase can also take sentences as complements to yield compound sentences of the form *Shìshí bìngfēi S* 'It is not the case that *S* (literally, The fact is not that *S*)', where *shìshí* and *bìngfēi* do not form a constituent, either.

[17] See, e.g., Martin (1992, 315–323), Song (1988, 71–166), Sohn (1999, 389–393), Lee and Chae (2000, Chap. 13), and Lee and Ramsey (2000, §6.4).

[18] Korean verbs do not have singular or plural forms, for Korean, like Chinese and a variety of other languages, has no grammatical number system. Thus the English translation of *ani-ta* in specific sentences can sometimes be 'is not' and sometimes 'are not'.

The adverb *an*, which modifies verbs, is a commonly used negative particle.[19] But it has serious limitations in forming negations.[20] By contrast, we can use the verb *ani*(-*ta*) to get negations of a wide variety of sentences.[21]

Like the English 'not' and the Chinese *bù*, the adverb *an* modifies verbs. And it is like *bù* in figuring before them. Consider (1)–(2):

(1) *Hana-ka* *ka-nta.*[22]
Hana-NOM go-DEC
'Hana is going.'

(2) *Hana-ka* *an* *ka-nta.*
Hana-NOM not go-DEC
'Hana is *not* going.'

By adding *an* before the verb *ka-* in (1), we can get its negation, (2), which amounts to 'Hana is *not* going.'[23] But we cannot use the adverb to get negations of sentential compounds, and adding it to some sentences (e.g., (4), the Korean cousin of 'All dogs are pretty') fails to yield their negations.

We cannot get a negation of (2) with *an*, for the adverb cannot modify the negative verb phrase *an ka-* '(do) not go'. Nor can we use it to get negations of conjunctions, such as (3):

(3) *Hana-ka* *ka-ko* *Toli-ka* *on-ta.*
Hana-NOM go-and Toli-NOM come-DEC
'Hana is going and Toli is coming.'

---

[19] The adverb is a short for *ani*, which is rather archaic. Note that the archaic adverb *ani* differs from the verb *ani-ta* (or its stem *ani-*), although they might have etymological relations. The negative verb might seem to derive from the combination of its affirmative counterpart *-i-ta* 'be' with the adverb *an* 'not', but **an i-ta* 'is not' (unlike, e.g., *an hada* '(do) not do') is not well-formed. Or it might derive from *ani hada* '(do) not do, is not', which has a secondary, copula use meaning *is not*, which is somewhat archaic (Martin *et al.* 1967, 1068). (More research would be necessary to verify etymological relations between the adverb and verb).

[20] The auxiliary verb cousin of *an*, *anh-ta* 'do not',is another major negation device. But it is not necessary discuss it for the present purpose, for it has roughly the same limitations as *an*.

[21] The verb *ani-ta*, like other Korean verbs, consists of the stem (*ani-*) and the suffix for basic forms of verbs (-*ta*). But I usually omit the suffix in giving Korean verbs, specifying only the stems (e.g., *ani-*).

[22] The declarative marker -(*n*)*ta* has two forms: -*nta* and -*ta*. They are suffixed to, e.g., *ka-* and *on-* in (1) and (3) below, respectively.

[23] In taking (2) to be a negation of (1), I make the usual assumption that they are contradictories. But this assumption is controversial for the possibility of *empty singular terms*. Some might hold that both are false if 'Hana' does not refer to anything. In this respect, (2) differs from the sentential negation of (1) given below, (10), which is its contradictory in the strict sense—(10) is true, if (1) is false because 'Hana' fails to refer. I ignore this potential difference between, e.g., (2) and (10) for the present purpose.

One might add it to the verbs in (3), but this does not yield its negation. Adding it to the verb *ka-* or *on-* yields conjunctions involving the negation of the left or right conjunct:

(3′) a. *Hana-ka an ka-ko Toli-ka on-ta.* 'Hana is *not* going and Toli is coming.'
b. *Hana-ka ka-ko Toli-ka an on-ta.* 'Hana is going and Toli is *not* coming.'

For the scope of the adverb is limited to the verb or predicate phrase modified by it. The same problem arises for even sentences with only one verb. Consider, e.g., a universal sentence:

(4) *Kay-nun motwu/ta yeyppu-ta.*
dog-TOP all/all pretty-DEC
'All dogs are pretty.'

Adding the adverb to (4) yields (5):

(5) *Kay-nun motwu/ta an yeyppu-ta.*
dog-TOP all/all *not* pretty-DEC
'All dogs are *not* pretty (viz., *No* dogs are pretty).'[24]

But (5) is not a negation of (4), for they are not contradictories (they can both be false).[25]

By contrast, we can use the verb *ani-ta* (or *ani-*) to get negations of a variety of declarative sentences, including (1)–(5). The verb is the negative counterpart of the suffix *-ita* (*or -i-ta*) 'be', which is used primarily as a *copula*,[26] as in (6):

[24] The Korean sentence is comparable to the English 'All dogs are *not* pretty' in syntax. Unlike the English sentence, however, it cannot be used to mean *Not all dogs are pretty.* (In English, it is not unusual to use, e.g., 'All that glitters is *not* gold' instead of '*Not* all that glitters is gold.') In both (4) and (5), the adverbial quantifier *motwu* and *ta* can be removed. Removing them yields a bare nominal (or so-called generic) sentence, which has the universal meaning (*all* dogs are pretty) for the topic/contrast suffix *-nun*—unless the suffix is taken to indicate a contrast between the dogs and some others (e.g., cats) prominent in the given context (see also note 35). (Note that the bare nominal sentences cannot be used to make existential statements.)

[25] The same problem arises for adding *an* to the Korean translation of the plural 'Hana and Toli are going/coming/pretty'.

[26] The suffix *-i-ta* 'be' consists of the stem *-i-* and the suffix for the basic form *-ta*. Like *ani-ta*, *-i-ta* has another use, which requires only one argument (see (8) below). Unlike the English 'be', however, the Korean *-i(-ta)* cannot be used for existence. Similarly, its negative counterpart *ani(-ta)* 'be not' cannot be used for nonexistence.

(6) *Hana-nun haksayng-i-ta.*
Hana-TOP student-be-DEC
'Hana *is* a student.'

The verb *ani-* is also used primarily as a copula, a *negative copula* (the Korean cousin of the Chinese *fēi* 'be not'), as in a negation of (6):

(7) *Hana-nun haksayng-i ani-ta.*
Hana-TOP student-NOM be.not-DEC
'Hana *is not* a student.'

In their copula uses, both *-i-ta* and *ani-ta* require two arguments: (a) the subject and (b) the complement. In (6) and (7), *Hana* 'Hana' is the subject and *haksayng* 'student' the complement.[27] But they also have *non-copular* uses, which involve only one argument. Consider, e.g., (8):

(8) *Ku kes-i ani-ta.*
That thing-NOM be.Not-DEC
'That (thing) is not [it, the case]'.[28]

This sentence has only one argument: *ku kes* 'that (thing)'.[29] Using (8) while referring to a statement someone made (e.g., with (4)) with the demonstrative phrase, one can state that it is not true (e.g., it is not true (or the case) that all dogs are pretty).[30]

And we can get negations of declarative sentences by replacing the demonstrative adjective *ku* 'that' in (8) with their *adjectival cousins*, i.e., the adjective clauses resulting from replacing their declarative endings (*-ta*, *-nta*) with suitable

[27] The nominative case marker *-i*, which is suffixed to *haksayng* in (7), is used for both subjects and (subject) complements.

[28] To distinguish the non-copular use of *ani-* from its copular use, I gloss the former with 'be.Not' (with the capital 'N'). In giving English translations of sentences involving it, it is necessary to add additional elements, e.g., 'the case', which I put in square brackets to indicate that there is no element matching it in the Korean sentences (see note 30).

[29] In colloquial speech, it is usual to use an abbreviation of the demonstrative phrase added with the nominative case suffix *-i* (i.e., *ku kes-i*): *kukey* 'that (nominative)'. The suffix *-ita*, too, has only one argument in the affirmative counterpart of (8) that features it (see (16) below).

[30] Note that although the English translation of (8) needs a complement for the copula 'is' (e.g., 'true', 'correct', 'the case'), the Korean *ani-* 'be not' has no content suggested by the English complement. (Nor does (8) have any other component providing the complementary content.) Note also that while the translation of *ku kes*, 'that (thing)', figures as the subject of the English translation, the Korean sentence might be considered a sentence without a subject. See Sect. 1.3, where I discuss how the non-copular use of *ani-* relates to its copular use.

forms of the *adjective clause suffix* $-(n)(u)n$ (ACS): *-n*, *-un*, *-nun*.[31] It would be useful to specify the general form of the resulting negations.

Let S be a declarative sentence, where the *declarative suffix* -$(n)ta$ figures as the sentence-final particle suffixed to the *last verb* of the sentence. Then let S^- be the sentential construction resulting from removing the suffix from S (and S_E an English translation of S). (Call S^- the *sentential base* of S.)[32] Then we can get the negation of S involving the non-copular use of *ani-* as follows:

(9) The (non-copular) *ani*-negation:
S^--$(n)(u)n$ *kes-i* *ani-ta.*
S^--ACS thing-NOM be.Not-DEC
'It is not [the case] that S_E.'[33]

For example, the following are the (non-copular) *ani*-negations of (1)–(4):

(10) *Hana-ka ka-nun kes-i ani-ta.*
Hana-NOM go-ACS thing-NOM be.Not-DEC
'It is not [the case] that Hana is going.'

(11) *Hana-ka an ka-nun kes-i ani-ta.*
Hana-NOM not go-ACS thing-NOM be.Not-DEC
'It is not [the case] that Hana is not going.'

---

31 See, e.g., (11)–(13) below, where the three forms are suffixed to *ka-*, *on-*, and *yeyppu-*. On the adjectival clause suffix, see, e.g., Lee and Ramsey (2000, 193–196). They call the suffix an "adnominal ending", and the adjective clause that includes it, which is used "as an element modifying a following noun phrase", an "adnominal" clause (2000, 193–196 & 203).

32 While adding the declarative suffix -$(n)ta$ to a sentential base S^- (e.g., *Hana-ka ka-*) yields a *declarative* sentence (e.g., (1)), adding an interrogative suffix (e.g., *-ni*) yields a (polar) *interrogative* sentence (e.g., *Hana-ka ka-ni*? 'Is Hana going?'). See the discussion of Korean interrogatives (e.g., (33)–(34)) in §2.2.

33 The *ani*-negation construction has two minor variants:

(a) The one where *kes-i* is replaced by the short form *key*.
(b) The one where it is replaced by *kes-un*, where the topic/contrast marker *-un* replaces the nominative marker *-i*.

(a) is a casual form; (b) can yield more natural instances in some cases. Martin (1992, 322) mentions (a) to note that one can use it to get double negations (e.g., (11)). While this is correct, it is worth adding that application of the *ani*-construction (and its variants) is not limited to negative sentences (see below).

(12)

Hana-ka	*ka-ko*	*Toli-ka*	*on-un*	*kes-i*	*ani-ta.*
Hana-NOM	go-and	Toli-NOM	come-ACS	thing-NOM	be.Not-DEC

'Not both Hana is going and Toli is coming.'[34]

(13)

Kay-ka	*motwu/ta*	*yeyppu-n*	*kes-i*	*ani-ta.*
dog-NOM	all/all	pretty-ACS	thing-NOM	be.Not-DEC

'Not all dogs are pretty.'[35]

The sentential base (S^-) of the adverb phrase modifying *kes* 'thing' in (10), for example, is *Hana-ka ka-*, which results from removing *-nun* from (1) (S). And (10), the *ani*-negation of (1), is its contradictory that in effect states that what the sentence states is not the case, for the adjective clause modifies *kes* to yield a noun phrase comparable to the English clause 'that Hana is going'.[36] The same holds, *mutatis mutandis*, for the other *ani*-negations, including (11)–(13).[37]

34 Note that neither the declarative nor the adjective clause suffix is added to the first verb *ka-* in (12). The declarative suffix -$(n)ta$ is a *sentence-final* particle and placed once in a declarative sentence as a suffix to the *last* verb. Similarly, the adjective clause suffix -$(n)(u)n$, which replaces the declarative suffix, is suffixed to the last verb of the sentential construction it governs.

35 Note that *kay* 'dog' is suffixed by the topic/contrast marker *-nun* in (4), but by the nominative case maker *-ka* in (13). While the topic/contrast marker can replace nominative markers (*-ka*, *-i*) in some sentences (e.g., universal sentences in the usual contexts), it is more natural to preserve *-ka* in some others (e.g., negations of universal sentences). Note also that replacing the nominative case marker *-i* suffixed to *kes* in (13) with the topic/contrast marker *-nun* yields a more natural sentence (see note 24).

36 So (10) is logically equivalent to (2), assuming that the latter is also a contradictory of (1) (see note 23).

37 (11) is a double negation of (1), amounting to $\sim\sim P$ (where P amounts to 'Hana is going'), (12) amounts to the symbolic $\sim (P \wedge Q)$ (where P and Q amount to 'Hana is going' and 'Toli is coming', respectively), and (13) a negation of the universal (4). Note that it is straightforward to get the *ani*-negations of disjunctions and conditionals. We can get the *disjunction* and *conditional* cousins of (3) simply by replacing the conjunction suffix *-ko* 'and' in (3) with the disjunction and conditional suffixes: *-kena* 'or' and *-myen* 'if ..., then ...'. (The conditional marker is attached to the antecedent and followed by the consequent (i.e., 'P-*myen* Q' for 'If P, then Q').) And the matching cousins of (12) are negations of those compounds, i.e., Korean counterparts of $\sim (P \vee Q)$ and $\sim (P \rightarrow Q)$. The *ani-* negations of some forms of *particular* sentences (e.g., Korean cousins of 'Some dogs are pretty') sound unnatural, but that might be because those partciular sentences are close to particular sentences involving *specific* indefinites (e.g., '*Certain* dogs (i.e., some *specific* dogs) are pretty'). Matching *particular* sentences of *other forms* and *existential* sentences (e.g., Korean cousins of 'There are dogs that are pretty') have unexceptionable *ani*-negations.

1.3 Grammaticalization of Korean Copulas

We can apply the non-copular *ani*-construction (9), '. . . *-nun kes-i ani-ta*', we have seen, to get negations of a wide range of sentences. Can the construction be considered a sentential negation device? I think the answer is yes. Like the English 'It is not the case that . . . ' and the Chinese '*Shìshí bìngfēi* . . . ', the Korean construction involves multiple components that do not clearly seem to make a constituent of sentences resulting from its application (e.g., (10)–(13)). Still, I think it differs from the English and Chinese cousins in being an idiomatic device designed for yielding negations involving a *grammaticalized* use of *ani-* that integrates the verb with the preceding part (*-nun kes-i*) by, so to speak, bleaching out key features of its original copula use.[38]

To see this, it would be useful to compare the non-copular *ani*-construction with its copular cousin:

(14) The copular *ani*-negation:

S-nun	*kes-un*	*sasil-i*	*ani-ta.*
S-ACS	thing-TOP	fact-NOM	be.not-DEC

'It is not a fact that S_E (literally, that S_E is not a fact).'[39]

The construction featured in this form, the copular *ani*-construction, where the verb *ani-* figures as a (negative) copula complemented with the noun *sasil* 'fact', is a close cousin of the English 'It is not the case that . . . ' and the Chinese '*Shìshí bìngfēi* . . . '.[40] And we can use the construction to get contradictories of declarative sentences, including (1)–(4). Applying it to, e.g., (1) yields (15):

(15)

Hana-ka	*ka-nta-nun*	*kes-un*	*sasil-i*	*ani-ta.*
Hana-NOM	go-DEC-ACS	thing-TOP	fact-NOM	be.not-DEC

'It is not a fact that Hana is going (literally, that Hana is going is not a fact).'

If so, can we consider the non-copular construction a short for the copular construction? I think not. Adding *sasil-i* before *ani-ta* in instances of the non-copular form (9) does not yield instances of the copular form (14), for the adjective clause suffix $-(n)(u)n$ is added to *entire* declarative sentences in the latter but only to their *sentential bases* resulting from removing the declarative suffix $-(n)ta$ in the former. Thus adding *sasil-i* to (10), for example, does not yield (15), but an ill-formed construction:

---

38 Some might take the same to hold for the English 'not that' and the Chinese *bùshì* 'not correct, not that'. But uses of the Korean construction are not limited to pragmatically constrained contexts (e.g., contrastive contexts).

39 Note that the adjective clause suffix *-nun* is to be added to *entire declarative sentences* (replacing *S*) in instances of this form.

40 And *sasil* 'fact' is a Sino-Korean noun stemming from the Chinese *Shìshí.*

(15′)**Hana-ka* *ka-nun* *kes-un* *sasil-i* *ani-ta.*
Hana-NOM go-ACS thing-TOP fact-NOM be.not-DEC
'*meant*: It is not a fact that Hana is going.'

The customary English translations of *ani*-negations might suggest that the verb *ani-* in them can be considered a copula with implicit complements. But *ani*-negations, we have seen, cannot be turned into matching sentences involving its copular use by adding complements. If so, how does the non-copula use of the verb relate to its copular use?

I think it might be taken to derive from omitting the *subject* rather than the *complement.* Consider (16)–(17):

(16) *Ku* *kes-i-ta.*
That thing-be-DEC
'That is [it, the case] (literally, [it, the case] is that).'[41]

(17) *Ku* *kes-i* *ani-ta.*
That thing-NOM be.Not-DEC
'That is not [it, the case] (literally, [it, the case] is not that).'[42] (=(8))

Both (16) and (17) (=(8)), which have *-i-ta* 'be [the case]' and *-ani-ta* 'be not [the case]', respectively, have only one argument: *ku kes* 'that (thing)'. But we can take them to have *implicit* subjects while taking *ku kes* to be the complement.[43] The missing subjects given implicitly in the contexts of their use might be expressions (e.g., nouns, pronouns) that refer to, e.g., what the speaker is looking for. Now, as noted above (Sect. 1.2), one can use (17) while using the demonstrative *ku kes* to refer to what someone states with declarative sentences (e.g., (1)). In such cases, what one states with (17) are the denials of the statements they make with the sentences (e.g., that it is not the case that Hana is going). Similarly, one can use (16) while referring to what is stated with, e.g., (1). In this case, what one states with (16) is the affirmation of the statement that Hana is going: it *is* the case that Hana is going. If so, can we take such

[41] In colloquial speech, *kes-i-* in (16) is usually contracted to *ke-*. (The same holds for the examples given below that includes *kes-i-*.) And *ku kes-i-* in (16) is usually contracted further to *kuke-* while the adverb *paro* 'exactly' is added, as in *Paro kuke-key* '(Exactly) *that*'s it.'

[42] In colloquial speech, *kes-i* in (17) is usually contracted to *key*. (The same holds for the examples given below that includes *kes-i-*.) And *ku kes-i-* in (17) is usually further contracted to *kukey*.

[43] Note that we cannot consider *ku kes* the subject of *-i-ta* in (16), for the suffix is added to the complement when used as a copula.

uses of (16) and (17) to involve implicit subjects? I think we can. We might take the subjects in such cases to refer to the *facts* in question in the contexts. In such cases, they are equivalent to sentences that have *sasil* 'fact' as the explicit subject:

(18) *Sasil-i* *ku* *kes-i-ta.*
fact-NOM that thing-be-DEC
'That is the fact (literally, the fact is that).'

(19) *Sasil-i* *ku* *kes-i* *ani-ta.*
fact-NOM that thing-NOM be.not-DEC
'That is not the fact (literally, the fact is not that).'

(18)–(19), where *sasil* 'fact' is the subject referring to the facts in question in the contexts of their use, are commonly used to state that another statement (i.e., the referent of *ku kes*) is true or not true (or does or does not state a fact). While using the demonstrative phrase to refer to the statement someone makes with, e.g., (1), one can use (19) to state that (1) does not state a fact.

Does this mean that *all* apparent cases of non-copula uses of *-i-ta* and *-ani-ta* are actually cases of their copula uses with implicit subjects? I think not. The implicit subject approach has essentially the same problem as taking the non-copular *ani*-negation form (9) to be a short for the copular *ani*-negation form (14). Adjective clauses matching declarative sentences (e.g., (1)) that can replace the demonstrative *ku* in (19) must be instances of *S-nun*, not instances of S^{-}-$(n)(u)n$, which figure in non-copular *ani*-negations. Replacing *ku* in (19) with the *S-nun* instance matching (1), for example, we can get (20):

(20) *Sasil-i* *Hana-ka* *ka-nta-nun* *kes-i* *ani-ta.*
fact-NOM Hana-NOM go-DEC-ACS thing-NOM be.not-DEC
'The fact is not that Hana is going.'[44]

But we cannot get (10) by removing *sasil-i* from (20) for the presence of the declarative *-nta* suffixed to the verb *ka-*. And removing the suffix from (20) yields an ill-formed construction.

I think the non-copular *ani*-construction derives from the *grammaticalization* of the copula *ani-* that involves a *re-analysis* of (19) and the like. Consider (21)–(22):

(21) *Sasil-un,* *ku* *kes-i-ta.*
fact-TOP, that thing-be-DEC
'As a matter of fact, that's [it, the case].'

[44] Notice that the construction in this sentence draws a close parallel with the Chinese *Shìshí bìngfēi* ... 'The fact is not (that) ...'.

(22) *Sasil-un,* *ku* *kes-i* *ani-ta.*
fact-TOP, that thing-NOM be.Not-DEC
'As a matter of fact, that's not [it, the case].'

(21) and (22) are closely related to (18) and (19), respectively, for the nominative suffix *-i* added to the subject *sasil* 'fact' in the latter can be replaced by the topic/contrast marker $-(n)un$. In (21) and (22), however, the topic/contrast phrase *sasil-un* is not considered the subject but an optional introductory phrase amounting to the English 'actually', 'as a matter of fact', 'to tell the truth', etc. This may be taken to result from a re-analysis of (18) and (19) (or their *-un* cousins), which results in a reading that takes *-i-ta* and *ani-ta* in them not to require two arguments. On this analysis, which recognizes non-copular uses of *-i-ta* and *ani-ta* that require only one argument, (16)–(17) are *complete* sentences that do not need another argument (explicit or implicit), and the introductory topic phrase can be optionally added to yield (21)–(22) and thus (18)–(19) (or their *-un* cousins).

Now, consider the results of replacing the demonstrative *ku* 'that' in (16)–(17) with instances of S^{-}-$(n)(u)n$, such as (23)–(24):

(23) *Hana-ka* *ka-nun* *kes-i-ta.*
Hana-NOM go-ACS thing-NOM be-DEC
'Hana is going; Hana is going; it is that Hana is going.'[45]

(24) *Hana-ka* *ka-nun* *kes-i* *ani-ta.*
Hana-NOM go-ACS thing-NOM be.Not-DEC
'[It is] not [the case] that Hana is going.'(= (10))

These sentences differ from, e.g., (15) and (20) in lacking the declarative marker *-nta* in *ka-nun* (cf., *ka-nta-nun*). In (15) and (20), which retain the marker, (1) is embedded in an *indirect quotation* clause, the instance of *S-nun kes* that amounts to the English 'that Hana is going'. By contrast, (1) does not figure as an embedded sentence in (23)–(24). But can the S^{-}-$(n)(u)n$ *kes* construction be taken to result from abbreviating the *S-nun kes* construction? I think not. I think it derives from a further re-analysis of the likes of (24) (and (23)) that integrates *ani-ta* (and *-i-ta*) with the preceding components: -$(n)(u)n$ *kes-i*.

[45] (23) is usually used instead of (1) for emphasis or contrast, usually on the subject. When the emphasis is on *ka-* 'go', the subject might be omitted in suitable contexts. (The underlines in the subject or predicate in the translations indicate the emphasis amounting to *-nun kes-i-*.)

To see this, it is useful to consider instances of the future-tense (or past-tense) cousins of S^{-}-$(n)(u)n$ *kes-i-ta*, such as (23′):

(23′) *Hana-ka* *ka-l* *kes-i-ta.*
Hana-NOM go-**ACS.FUTURE** thing-be-DEC
'Hana **will** go.'(=(10))

(23′) results from (23) by replacing the plain adjective clause suffix -$(n)(u)n$ with its future tense cousin: *-l*. But (23′) is *not* taken to have a *clause* embedding (1). As the translation given above ('Hana *will* go') suggests, it is considered just the future tense cousin of (1) that includes *-l kes-i* as a future tense marker.[46] On this analysis, *kes-i-* (which includes a non-copula use of *-i*(*-ta*)) is integrated with *-l* to form a suffix phrase indicating the future tense, and (23′) results from (1) by inserting the phrase between the verb *ka-* and the declarative suffix -(n)*ta*.[47] And we can give the same analysis for the plain cousin of (23′): in (23), -$(n)(u)n$ *kes-i-* is a suffix phrase for emphasis or contrast added to a verb before the declarative marker -(n)*ta* and combines with this suffix to form the sentence-final suffix phrase -$(n)(u)n$ *kes-i-ta.*

The above analysis of the likes of (23) and (23′) is their standard analysis in Korean grammar. So Baek (1999, 86), for example, lists -$(n)(u)n$ *kes-i-ta* (and its past- and future-tense cousins) among the Korean sentence-final suffix phrases, and she says that -$(n)(u)n$ (or the like), *-kes*, and *-i-ta* "combine to play the role of a [suffixed] ending of verbs" (1999, 86).[48] We can take the same to hold, *mutatis mutandis*, for the negative counterpart of -$(n)(u)n$ *kes-i-ta* in non-copular *ani*-negations: -$(n)(u)n$ *kes-i ani-ta.* This is a sentence-final suffix phrase resulting from adding the declarative suffix -(n)*ta* to the suffix phrase -$(n)(u)n$ *kes-i ani-*,

---

[46] Note that '**Hana-ka ka-**l**-ta*' is ill-formed. So *-l* by itself is not sufficient for a future tense sentence and the addition of *kes-i* is essential.

[47] The same holds, *mutatis mutandis*, for the past-tense cousin of, e.g., (23), which results from replacing *ka-nun* with *ka-n*.

[48] See also Sohn (1999), who says that the verb *-i-ta* 'be' in sentences like (23′) "have become so grammaticalized with the preceding elements that they have lost their ability to carry their own subject" (1999, 292). Note that further progress in grammaticalizing the *-i-ta* in the likes of (23) is made with the contraction of *kes-i* to *ke*, as in (23″):

(23″) *Hana-ka* *ka-l* *ke-ta.*
Hana-NOM go-ACS.FUTURE (thing+be)-DEC
'Hana will go.'

The contraction *ke* is just a part of *kes* and has no apparent trace of the verb -i-. Moreover, it is not unusual to merge *-l* and *ke*(*-ta*) to *-l-ke*(*-ta*) (without space) although the standard Korean orthography requires separation of *-l* and *ke*(*-ta*), as in (23″), to reflect the origin of *ke*.

and (24), for example, results from suffixing the verb *ka-* with the phrase. On this analysis, *ani-* is integrated with the preceding -$(n)(u)n$ *kes-i* (in short, -$(n)(u)n$ *key*) to form a suffix phrase serving as a sentential negation device.[49]

On the above account of the non-copular uses of *-i-ta* and *ani-ta* in the likes of (23)–(24), they lack essential features of copulas for the grammaticalization resulting in their uses in the sentences. The account distinguishes their uses in such sentences not only from (a) their original *copula* uses but also from (b) their uses in the likes of (16)–(17). In these intermediate uses, (b), they lack features essential to *copula* uses, (a), but still retain features essential for *verbs*: they require an argument, and *ku kes* is their subject in, e.g., (16)–(17). In (23)–(24), however, they cease to be even verbs to retain only the features of *affirmation* and *negation* in their copula ancestors. These features are highlighted in their free-standing uses for *affirmation* and *denial*, that is, *agreement* and *disagreement* with what someone said. The Korean cousins of 'Yes' and 'No' are *Yey* and *Ani-ta* (or the polite *Ani-yo*), respectively. *Ani-ta* and *Ani-yo* derive from the plain and polite declarative forms of *ani-*, namely, the results of suffixing it with the plain declarative -$(n)ta$ and the polite declarative *-yo*, respectively. *Yey* is related to a polite declarative form of *-i-ta*: *-i-e.yo* (3 syllables). Its 2 syllable contraction is *-yey.yo*, and we can get *Yey* from this by removing *yo*. So I think both *Yey* and *Ani-ta* (or *Ani-yo*) derive from the non-copula uses of *-i-* and *ani-* for *affirmation* and *negation*.

2 Disjunction and Choice Question

The English counterparts of *disjunction* symbols of symbolic languages (e.g., $\vee$, A) are 'or' and its variants (e.g., 'either ... or ...'). And 'or' can be used in both declarative and interrogative sentences, such as (25)–(26):

(25) I like coffee *or* tea.

(26) Do you like coffee *or* tea?

But (26) is ambiguous, as noted by, *inter alia*, Jespersen (1992, 303f; 1933, 306) and Chao (1948, 58f; 1955, 35). One can use it to mean either of two questions:

(27) a. Do you like (at least) one of coffee and tea?
b. Which of coffee and tea do you like?

49 Japanese has a counterpart of -$(n)(u)n$ *kes-i ani-ta: . . . -no dewa-na-i* 'It is not the case that . . .', where *dewa-* 'be' is a copula and *-na* 'not' a negative particle (Masuoka and Takubo 1989, §III.7.7–10; Imani 2020, 528ff). It is, I think, a sentential negation device deriving from the copula *dewa* and the negative marker *-na*.

(27a) is a *yes/no* or *polar question*, which one can answer with 'Yes' or 'No'. By contrast, (27b) is an *alternative* or *choice question*, which one cannot answer with 'Yes' or 'No'.[50] The question relates to two alternatives:

(28) a. You like coffee.
b. You like tea.

While presenting the alternatives, those who ask the question request the addressees to choose one of them that holds.[51] Chao (1955, 34–36; 1968, 265f) gives an interesting analysis of Chinese *choice interrogatives* (i.e., interrogative sentences one can use to ask choice questions). It is an insightful analysis we can extend for a variety of languages, including Korean (§2.2), Japanese (§2.3), and English (§2.4).

2.1 Chao's Analysis of Chinese Choice Interrogatives

The English interrogative (26), as noted above, is ambiguous. But one can disambiguate it with intonation in spoken language (see, e.g., TfCS (U.d.)). And replacing the plain 'or' in the sentence with the emphatic 'either ... or ...' yields an unambiguous cousin, an interrogative one can use for the polar question but not for the choice question (Haspelmath 2007, 26). But English might seem not to have a straightforward variant of (26), one with no question word (e.g., 'which'), that can be used *only* for a choice question.[52] Some languages (e.g., Chinese, Japanese, Korean, Basque), by contrast, have unambiguous choice interrogatives (without question words), which one cannot use for polar questions.[53]

Chinese (Mandarin) has two groups of expressions figuring in usual Chinese counterparts of English phrases involving 'or' and the like:

(a) *huòshì*, *huòzhě*, *huò*, etc.

[50] Chao, I think misleadingly, calls choice questions "disjunctive questions" (1948, 58). See also, e.g., Jespersen (1933, 304) and Haspelmath (2007, 26).

[51] The request assumes that at least one of the alternatives holds (i.e., the answer to the matching polar question, (27a), is yes).

[52] I do not think this is correct. For example, 'Do you like coffee, or do you like tea?' can be used only for a choice question (see (50b) in §2.4). The same holds for "Did he say that, or did he not (say that)?" (Jespersen 1924, 304), "Do you want me to help you, or do you want to do it by yourself" (Li and Thompson 1981, 654; Haspelmath 2007, 4), and "Do you want tea or do you want coffee?" (Haspelmath 2007, 25). (The latter two are given as translations of Chinese and Basque examples of unambiguous choice interrogatives).

[53] In addition to Chao (1948; 1955; 1968), see, e.g., Li and Thomson (1981, 653f), Erlewine (2014), Uegaki (2014b), Beck and Kim (2006, 171), Lee (2003, 356f), Haspelmath (2007, 25f), and Saltarelli (1988, 84) for discussions of Chinese, Japanese, Korean, and Basque choice interrogatives. See, e.g., Biezma and Rawlins (2015) for a survey of accounts of choice questions.

(b) *háishì*

While *huòshì* and the like can be used in both declaratives and interrogatives, Li and Thomson say that *háishì* can be "used only in questions" (1981, 653f).[54] And Chao, who notes that English interrogatives involving the plain 'or' (e.g., 'Will you eat rice or noodles') have the polar/choice ambiguity, says that one can use *huòshì* to translate 'or' when it figures in polar interrogatives, and *háishì* for choice interrogatives (1948, 58f).[55] Now, consider Chinese counterparts of (26) involving *huòshì* and *háishì*:

(29) *Nǐ xǐhuan kāfei huòshì chá?*
you like coffee <huòshì> tea
'Do you like (either) coffee or tea?' [polar]

(30) *Nǐ xǐhuan kāfēi háishì chá?*
you like coffee <háishì> tea
'Do you like coffee, or [do you like] tea?' [choice][56]

While (29) can be used for a polar question (viz., question (27a)), (30) cannot. Unlike the English (26), (30) does not have a polar/choice ambiguity and can be used *only* for a choice question (viz., question (27b)).[57]

So it is usual to take *háishì* to be for a special kind of *disjunction*, a *coordinator* amounting to the use of the English 'or' for choice questions, while taking *huòshì* (and the like) to amount to the use of 'or' in declaratives and polar interrogatives. For example, Haspelmath, who divides disjunctions into "standard" and "interrogative" ones, holds that *háishì* is a coordinator for *interrogative* disjunction while the others are coordinators for *standard* disjunction (2007, 3f).

[54] Li and Thomson list both among "adverbial backward-linking elements" (1981, 653).

[55] Li and Thomson take *háishì* to match the *exclusive* use of 'or' while taking *huòshì* and the like to match the *inclusive* use (1981, 653f), but the example they give of a sentence involving *háishì* is an unambiguous choice interrogative, for which they give the translation "Do you want me to help you, or do you want to do it by yourself" (1981, 654). (See also Haspelmath (2007, 4), who quotes their examples involving *háishì* and *huòzhě*.) They do not explain how the putative exclusive disjunction reading of *háishì* relates to the choice question reading. But I see no reason that exclusive disjunctions might not be used for polar questions. One might well ask 'Do you *either* want me to help you *or* want to do it by yourself?', while taking 'either ... or' to be for exclusive disjunction.

[56] Erlewine (2014, 231) gives essentially the same examples involving *huòshì* and *háishì*.

[57] Erlewine (2014, 231) suggests that the likes of (29) cannot be used for choice questions (cf., note 25 in Erlewine (2014, 232)). But J. Yan and J. Hung think that they do have the polar/choice ambiguity, while agreeing that the likes of (30) do not (private conversation).

And Chao divides "the 'or'-words in Chinese" into two kinds to call *háishì* "a disjunctive *'or'* (the 'or' of 'whether or')" and, e.g., *huáshì* "an alternative *'or'* (the 'or' of 'either or')" (1968, 265).[58] While presenting this account, which assumes that *háishì* amounts to the English 'or' in choice interrogatives, however, he holds that the assumption is "all right as a rule of thumb for translation purposes", but "quite misleading as grammatical analysis" (1955, 35). And he gives an interesting analysis of Chinese choice interrogatives.

In English and many other languages, it is usual to use *coordinators* (e.g., 'and', 'or') in coordinate compounds to connect phrases or other terms of coordination. But one can just *juxtapose* coordinated parts *without* a coordinator, as in the Latin *Veni, vidi, vici* 'I came, I saw, I conquered.' While such coordination, *asyndetic* coordination, is rather rare in English,[59] it is used frequently in Chinese, as Chao (1948, 37; 1955, 34f; 1968, 262ff) explains.[60] He says, "*Coordination* in Chinese is expressed by mere juxtaposition" (1948, 37, original italics). And he takes Chinese *choice interrogatives* to result from "simple juxtaposition" of two or more *polar interrogatives*, i.e., interrogatives one can use for polar questions. In choice interrogatives, on his analysis, *háishì* is not a coordinator connecting phrases or clauses for alternatives, but an optional adverbial element added to some of the polar interrogatives that yield choice interrogatives via mere juxtaposition (1955, 35f; 1968, 266f).

Consider the examples of Chinese choice interrogatives Chao uses to explain his analysis (1955, 35f):

(31) *Nǐ chī fàn háishì chī miàn?*
you eat rice <*háishì*> eat noodles
'Do/Will you eat rice, or eat noodles?' [choice]

(32) *Nǐ chī fàn chī miàn?*
you eat rice eat noodles
'Do/Will you eat rice, or eat noodles?' [choice]

In (31), *háishì* might seem to be a coordinator conjoining the clauses indicating two alternatives:

(a) (*Nǐ*) *chī fàn* '(You do/will) eat rice'
(b) (*Nǐ*) *chī miàn* '(You do/will) eat noodles'

[58] See also Chao (1948, 145). See also the account of Cantonese counterparts of *huòshì* and *háishì* in Matthews and Yip (1994/2011, §16.1.4 & §17.2).

[59] See, e.g., Greenbaum (1996, 532) for English examples of asyndetic coordination.

[60] See also, e.g., Li and Thompson (1981, 631) and Matthews and Yip (1994/2011, §16.1).

Unlike the 'or' in the English translation, however, *háishì* is not required in (31). Removing it yields (32), and we can use this to ask the same question. In (32), however, the two clauses are juxtaposed without a coordinator, which is sufficient for coordination in Chinese as noted above. So Chao holds that "a disjunctive question in Chinese is grammatically a conjunction, which usually takes the form of *simple juxtaposition*" (1955, 36, my italics).[61] To the simple juxtaposition (e.g., (32)), he holds, *háishì* (and the shorter *shì*) can be optionally added to mark the juxtaposed clauses: "before the terms of the grammatical conjunction one can add optionally *sh* [in pinyin, *shì*] . . . or *hairsh* [in pinyin, *háishì*]" (1955, 35, pinyin added). Listing various ways to add the optional elements, he says that adding *háishì* once as in (31) is "the favorite (most frequent)" because it "achieves the strongest effect with a minimum of effort", which leads to "the common practice" of considering it a coordinator comparable to 'or' (1955, 36).

I think this is an insightful analysis of choice interrogatives that lays out the basic components of choice interrogatives. On the analysis, they need not involve coordinators connecting clauses specifying the alternatives in question. But some might object that the Chinese *háishì* is an *essential* component of Chinese choice interrogatives although it differs from the English 'or' in that it can be *implicit*, as in (32), which might be considered a short for (31). To address this issue, it would be useful to consider languages with choice interrogatives that do not include even *apparent* coordinators. Korean and Japanese, I think, are such languages.

2.2 Korean Choice Interrogatives

Korean has many coordinators, including those amounting to 'and' and 'or'. And the language has several coordinators for *disjunction*:

(a) suffix: *-na*, *-kena*
(b) non-suffix: *ttonun*, *hok.un*

But none of them is used for choice interrogatives. They involve *mere juxtaposition* or *asyndetic coordination* of two or more polar interrogatives. So Korean has two unambiguous counterparts of the English interrogative (26):

(33) *Ne khephi-na cha coh.aha-ni?*
you coffee-or tea like-INT
'Do you like (either) coffee or tea?'[62] [polar]

[61] Chao uses "conjunction" to mean (*the result of*) *conjoining* (i.e., for coordinate compounds in general), not specifically for coordinate compounds involving the likes of 'and'. He also says, "the Chinese way of asking ... a [choice] question is to treat it as a grammatical conjunction of two coordinate terms . . . . The simplest [such] way . . . is to use co-ordination by [mere] juxtaposition" (1955, 36).

[62] The non-suffixal disjunction markers *ttonun* and *hok.un* can replace *-na* or be added after the suffix.

(34) *Ne* *khephi* *coh.aha-ni,* *cha* *coh.aha-ni?*
you coffee like-INT, tea like-INT
'Do you like coffee, or [do you] like tea?'[63] [choice]

(33) is a *polar interrogative* (i.e., an interrogative for a polar question), which one can use to ask (27a), and (34) a *choice* interrogative to use for (27b). They cannot be used for the other questions.

Note that (33) has a coordinator: the disjunction marker *-na* suffixed to the noun *khephi* 'coffee'. The same holds for the Korean polar interrogative matching 'Do you (either) *like coffee* or *like tea*?':

(35) *Ne* *khephi* *coh.aha-kena* *cha* *coh.aha-ni?*
you coffee like-or tea like-INT
'Do you (either) like coffee or like tea?'[64] [polar]

This sentence has the disjunction marker *-kena*, which is suffixed to verbs to conjoin predicate phrases (e.g., *khephi coh.aha(-ta)* 'like coffee'). By contrast, the choice interrogative (34) has no coordinator. (34) involves *mere juxtaposition* or *asyndetic syndetic* coordination of two polar interrogatives:

(36) a. *Ne* *khephi* *coh.aha-ni?*
you coffee like-INT
'Do you like coffee?' [polar]

b. *Ne* *cha* *coh.aha-ni?*
you tea like-INT
'Do you like tea?' [polar]

The particle *-ni* suffixed to the verb *coh.aha-* in these sentences is an *interrogative marker*.

Now, removing the interrogative particle *-ni* from polar interrogatives (e.g., (36a)–(36b)) yields the *sentential bases* they share with their declarative cousins, such as (37a)–(37b):

---

63 The non-suffixal disjunction markers *ttonun* and *hok.un* can neither replace nor be added to the first *-ni*.

64 The non-suffixal disjunction markers *ttonun* and *hok.un* cannot replace *-kena* but can be added after the suffix. Note that the choice interrogative cousin of (35) is the same as (34).

(37) a. *Ne* *khephi* *coh.aha-.*
you coffee like-
'Roughly, you like- coffee.' [sentential base]

b. *Ne* *cha* *coh.aha-.*
you tea like-
'Roughly, you like- tea.'[65] [sentential base]

While declarative sentences result from adding the declarative marker –(*n*)*ta* to sentential bases (e.g., (37a)–(37b)),[66] suffixing the interrogative marker *-ni* to them yields interrogative sentences (e.g., (36a)–(36b)), where *-ni* is suffixed only to the last verbs to take the sentence-final positions. Such interrogatives are *polar* interrogatives. Thus (36a)–(36b) are polar interrogatives. The same holds for (33), which differs from them only in having the disjunction *khephi-na cha* 'coffee or tea' instead of the simple *khephi* 'coffee' or *cha* 'tea'.[67] *Choice* interrogatives, by contrast, do not result from suffixing the interrogative marker *-ni* to some sentential bases. They result from combining two (or more) polar interrogatives (e.g., (36a)–(36b)) by mere juxtaposition, as in (34). Thus they have two or more occurrences of interrogative markers (e.g., *-ni*).

The juxtaposition involves no coordinator conjoining the juxtaposed parts. (34), for example, has no counterpart of 'or', 'and', or any other coordinator. Moreover, we cannot even *add* a disjunction marker. Adding the suffix *-na* 'or' for coordination of nouns to *khephi* 'coffee' or *cha* 'tea' yields an ill-formed construction. The same holds for adding the disjunction *-kena* for coordination of predicates before or after the interrogative *-ni* in its first part (i.e., the one before the comma), and replacing *-ni* in the part with *-kena* yields an unambiguous *polar* interrogative: (35).[68]

While we cannot add coordinators to Korean choice interrogatives, we can add some optional adverbs: *ani-myen* 'otherwise, if not (so)', *hoksi* 'perhaps', *chalali* 'rather', etc.[69] We can add these to the second part of (34) as follows:

(38) *Ne* *khephi* *coh.aha-ni,* *ani-myen* *cha* *coh.aha-ni?*
you coffee like-INT, if.not.so tea like-INT
'Do you like coffee, or else [do you] like tea?' [choice]

[65] In glossaries and English translations, the hyphen '-' is added to English verbs matching Korean verbs to indicate that the sentential bases involve the verbs themselves (or their stems), not their finite forms.

[66] See the discussion of sentential bases in §1.2 (esp., note 32).

[67] The disjunction suffix *-na* is not related to the interrogative suffix *-ni*.

[68] The non-suffixal coordinators for disjunction (*ttonun*, *hok.un*) cannot be added to the choice interrogative (34), either (see note 63).

[69] *Ani-myen* 'if not (so)' results from adding the conditional suffix *-myen* 'if' to *ani-* 'be not, not'. *Kulehchi anh.umyen* 'otherwise, if not so' is synonymous with *ani-myen* but more formal.

(39) *Ne* *khephi* *coh.aha-ni,* *ani-myen* *hoksi* *cha* *coh.aha-ni?*
you coffee like-INT, if.not.so perhaps tea like-INT
'Do you like coffee, or else [do you] perhaps like tea?' [choice]

We can add *ani-myen* 'if not (so), otherwise, or (else)', as in (38); with this added, we can further add *hoksi* 'perhaps', as in (39).[70] But neither of them is a coordinator. They are rather adverbs added to the second part of (34): (36b). Clearly, *hoksi* 'perhaps' is not a coordinator. Neither is *ani-myen* 'if not (so)'. To see this, consider the two parts of, e.g., (38):

(40) a. *Ne* *khephi* *coh.aha-ni?*
you coffee like-INT
'Do you like coffee?' [=(36a), polar]

b. *Ani-myen* *(ne)* *cha* *coh.aha-ni?*
if.not.so (you) tea like-INT
'Or (else) do you like tea?' [polar]

Both are complete sentences one can use to ask polar questions. Consider a discourse where (40b) follows (40a) to take it as an antecedent:

(41) *Ne* *khephi* *coh.aha-ni?* *Ani-myen* *cha* *coh.aha-ni?*
you coffee like-INT? otherwise tea like-INT?
'Do you like coffee? Or (else) do you like tea?'

As is clear from this discourse, *ani-myen* does not figure in (40b) as a coordinator conjoining two clauses for a coordinate construction, but as an adverbial element amounting to 'if not (so)', 'otherwise', or 'or (else)', which can be replaced in the English translation of (41) by the subordinate clause '*if* you do *not* like coffee'.[71] And (40b) is used in the discourse to ask a *conditional* polar question: Do you like tea if you don't like coffee? And we can just juxtapose (40b) after

[70] We can add *chalali* 'rather' like *hoksi* 'perhaps' to *ani-myen* in the Korean translation of, e.g., 'Would you like to have coffee, or (else) rather (to have) tea'.

[71] This subordinate clause is related to the antecedent interrogative: 'Do you like coffee?' This does not mean that the clause (or 'otherwise' or 'if not (so)') has an underlying element that conjoins 'Do you like tea?' to the antecedent *interrogative*. For 'you do *not* like coffee' is not the negative of the interrogative, but of its declarative cousin: 'You like coffee' (see the discussion of (49h)–(49j) in §2.4). The same holds, *mutatis mutandis*, for their Korean counterparts.

(40a) to get a choice interrogative: (38). So (38) is an asyndeton, involving mere juxtaposition of two polar interrogatives.[72]

2.3 Japanese Choice Interrogatives

Japanese choice interrogatives have essentially the same structure as Korean ones.[73] So it is straightforward to apply Chao's analysis of Chinese choice interrogatives to their Japanese counterparts. And it is usual in Japanese grammar to take them to be asyndetons, resulting from mere juxtaposition of two or more polar interrogatives. Masuoka and Takubo say that we can get Japanese choice interrogatives "by combining two or more yes/no interrogatives", and that they "usually" have the asyndetic form "A-*desu-ka*, B-*desu-ka*", where *-ka* is a sentence-final interrogative marker suffixed to verbs, e.g., the copula *desu* 'be' (1992, §III.7.2).[74] And Hinds says that "Alternative questions in Japanese . . . take the form of two complete questions", and that the asyndetic form is "the *canonical* form of alternative questions" (2003, 23, my italics).

Consider, e.g., (42):

(42)	*Taro-ga*	*koohii-o*	*non-da-ka,*	*ocha-o*	*non-da-ka?*
	Taro-NOM	coffee-ACC	drink-PAST-INT,	tea-ACC	drink-PAST-INT

'Did Taro drink coffee, or did [he] drink tea?'[75] [choice]

This is a choice interrogative resulting from juxtaposing two polar ones:

(43) a.	*Taro-ga*	*koohii-o*	*non-da-ka?*
	Taro-NOM	coffee-ACC	drink-PAST-INT

'Did Taro drink coffee?' [polar]

b.	(*Taro-ga*)	*ocha-o*	*non-da-ka?*
	Taro-NOM	tea-ACC	drink-PAST-INT

'Did Taro/[he] drink tea?' [polar]

[72] I think, for essentially the same reason, that English choice interrogatives (e.g., the interrogative version of (26)) also result from mere juxtaposition of two or more polar interrogatives (e.g., 'Do you like coffee?' and 'Or (else) do you like tea?'). See §2.4.

[73] See, e.g., Hinds (2003, 23ff) and Martin (1962, 288ff; 2004, 924ff) for examples and discussions of Japanese choice interrogatives.

[74] The Japanese interrogative suffixes *-ka* and *-no* differ from their Korean cousins in that they are usually *just added* to the last verbs of declarative sentences (e.g., *desu* 'be').

[75] This example is from Uegaki (2014a, 48ff; 2014b, 257ff), but I add the comma to make clear the separation between the two parts.

We can get (42) by adding (43b) to (43a) while removing the parenthetical subject clause, which (43b) shares with (43a).

Now, Japanese has an expression often added to choice interrogatives: *sore-tomo* 'or (else)'. We can add it to (42) as follows:

(44) *Taro-ga* *koohii-o* *non-da-ka,* *sore-tomo* *ocha-o* *non-da-ka?*
Taro-NOM coffee-ACC drink-PAST-INT, or.else tea-ACC drink-PAST-INT
'Did Taro drink coffee, or (else) did [he] drink tea?' [polar]

In this sentence, *sore-tomo* is placed between the parts matching (43a)–(43b). Uegaki (2014a; 2014b) takes this to mean that Japanese choice questions "are always disjoined PolQs", namely, *disjunctions* of two or more polar questions (2014b, 252). He holds that *sore-tomo* is "a specialized disjunction marker that is used to disjoin PolQs", and that it is manifest in (44) but "covert" in (42) (2014a, 52; 2014b, 258).[76] But he gives no reason to assume that (42), which involves no apparent coordinator, is not asyndetic like Korean choice interrogatives (e.g., 34) but involves a covert coordinator.[77] Would he still be right to take *sore-tomo* to be a coordinator conjoining polar interrogatives? I think not.

Those who hold the usual analysis of Japanese choice interrogatives deny that *sore-tomo* is a coordinator. Hinds, who takes the asyndetic form to be canonical, adds that *sore-tomo* "frequently separates the alternative forms" (i.e., the component polar interrogatives, e.g., (43a)–(43b)), and says that the phrase "functions to set off the alternatives" (2003, 23). And Martin explains its role in choice interrogatives as follows:

> Sometimes Japanese add the expression **sore tó mo** '[also with that =] or else' like an adverb before the last alternative suggested in an alternative question . . . . The expression **sore tó mo** only emphasizes the fact that you are representing alternatives—the sentence would mean just about the same without the expression. (Martin 1962, 231)

On their account, *sore-tomo* is not a coordinator that conjoins (43a) and (43b) in (44), but an adverbial phrase that rather helps to separate them in the asyndetic compound by marking the beginning of its second part.

To see this, it is useful to note that *sore-tomo* can be added to polar interrogatives as in (43c):

[76] He says that *sore-tomo* is "optionally covert in an embedded environment" (2014b, 258). But this does not explain its absence in the choice interrogative (42), which has no embedded environment.

[77] He argues that the first *-ka* in (42) is not a coordinator conjoining predicate phrases (*koohii-o non-da* and *ocha-o non-da*), and then simply *assumes* that the sentence must have a coordinator to conclude that it has a covert coordinator.

(43) c. *Sore-tomo* (*Taro-ga*) *ocha-o* *non-da-ka?*
Or.else Taro-NOM tea-ACC drink-PAST-INT
'Or (else) did Taro/[he] drink tea?'[polar]

In this sentence, *sore-tomo* is not a coordinator conjoining (43c) with antecedent interrogatives, but an adverbial phrase added to (43b) as a short for a subordinate clause amounting to 'If not *so*', which has an explicit part (the italicized 'so') that relates to an antecedent. And we can use (43c) after (43a) in a discourse:

(45) *Taro-ga* *koohii-o* *non-da-ka?* *Sore-tomo* *ocha-o* *non-da-ka?*
Taro-NOM coffee-ACC drink-PAST-INT? Or.else tea-ACC drink-PAST-INT
'Did Taro drink coffee? Or (else) did [he] drink tea?'[78] [polar]

So (44) can be considered a mere juxtaposition of two interrogatives: (43a) and (43c) (instead of (43b)).

The Japanese *sore-tomo* 'or (else)', we have seen, has essentially the same function in choice interrogatives as the Korean *ani-myen* 'if not (so)'. Unlike *ani-myen*, however, *sore-tomo* does not derive from parts amounting to 'if' and 'not'.[79] In morphology, it is close to *sore-temo* 'even so, still, nevertheless'; *sore* amounts to 'that', and *-tomo*, like *-temo*, amounts to 'even if', 'however', etc. So one might take it to amount originally to 'in spite of that', 'even so', 'still', 'nevertheless', 'nonetheless', etc. And it can indeed be used with this meaning in Japanese translations of, e.g., the following:

(a) We are doomed if we do not study tonight. *Even so*, would you insist that we go out to see a movie?
(b) I would suggest that we go out to see a movie. Would you like to do that? Or would you *still* like to stay at home?

In translating these discourses into Japanese, we can replace the italicized 'even so' and 'still' with *sore-tomo*.[80] Nevertheless, it is usually used in modern Japanese essentially as an introductory marker for alternatives presented in choice interrogatives or the like. So it is usual to translate it by 'or' or 'or else', and a Japanese-English dictionary explains its use as follows:

[78] See the examples in the entry for *sore-tomo* in Japan Foundation (2004, 467).

[79] Much of the subsequent discussion of *sore-tomo* draws on Youichi Matusaka's descriptions of its meaning and use in private exchange, for which I wish to thank him. Still, I am solely responsible for any error in the discussion, which includes my rephrasing and interpretation of his descriptions.

[80] Although I have made some stylistic changes, the two examples are Matsusaka's (private exchange).

soretomo ⟦*conj*⟧ **or, or else** . . . . *Used when presenting a choice of two items. Usually used in the pattern "∼ *ka? Soretomo,* ∼ *ka?*" (Japan Foundation 2004, 467)[81]

In this use, *sore-tomo* does not suggest any *presumption* that the alternatives presented in the antecedent polar interrogatives are what one might usually expect. Despite *sore-tomo* in (43c), those who use the interrogative in (44)–(45) might well assume that the addresses are as likely to drink tea as to drink coffee. This *presumption-free* use of *sore-tomo*, I think, might have been derived from the literal and probably earlier use.

We can imagine a process in which *sore-tomo* might shed the presumption for the antecedent alternatives in its literal meaning to develop the presumption-free use. Consider (46a)–(46d):

(46) a. (If you like, we can go out. So . . .) Would you like to go out to see a movie? *In spite of that*, would you like to stay at home?
b. (If you like, we can go out. So . . .) Would you like to go out to see a movie? (Or) Would you *still* stay at home?
c. (If you like, we can go out. So . . .) Would you like to go out to see a movie? (Or) Would you *rather* stay at home?
d. (If you like, we can go out. So . . .) Would you like to go out to see a movie? (Or) Would you *instead* stay at home and watch TV?

In (46a)–(46b), 'in spite of that' and 'still' suggest the presumption that the alternative presented in the antecedent interrogative might be preferable. But no such presumption is suggested by 'rather' or 'instead' in (46c)–(46d). So *sore-tomo* might be taken to amount to 'rather' or 'instead' in its usual use as a marker of later parts of choice interrogatives. If so, (44) is tantamount to (47):

(47) Did Taro drink coffee, or is it *rather*/*instead* the case that he drank tea?

So I think the *sore-tomo* in (44) is like the Korean *chalali* 'rather' in meaning than the Korean *ani-myen* 'if not (so)' (or the English 'or'), *except* that it is used like the latter to introduce the clause for the second alternative, (43b), by being placed in front of it because it is not accompanied by a Japanese counterpart of *ani-myen* (or 'or').

On this analysis, the choice interrogative (45) has no counterpart of the English 'or'. But (47) has not much difference in meaning from usual translations of (45):

(48) a. Did Taro drink coffee, or did he drink tea? [choice]

[81] Note that the form given in this entry is one for a *discourse* involving polar interrogatives with *sore-tomo* (e.g., (45)), not one for *choice interrogatives* (e.g., (44)).

b. Did Taro drink coffee or (drink) tea? [choice]

This might lead some to identify *sore-tomo* with the English 'or' to consider it a disjunction marker. But the parallel between them (and the Korean *ani-myen*) arises just from the fact that they all can be used as *customary markers* (mandatory or not) of some parts of choice interrogatives (e.g., (43c)). This, we have seen, does not mean that they have the same meaning, nor that they are all coordinators, let alone those for disjunction. We can now see that even the English 'or' is not used as a coordinator in choice interrogatives.

2.4 English Choice Interrogatives

Usual analyses of choice interrogatives assume that English choice interrogatives involve *syndetic* coordination, where 'or' figures as a kind of *disjunction* marker. This ignores the connection that its use for choice interrogatives has to 'or else', 'else', 'otherwise', 'if not (so)', etc. By attending to the connection, we can see that English choice interrogatives also involve mere juxtaposition, albeit with *superficial* resemblance to syndetic coordination.

The English 'or' has various uses, including some that cannot be considered uses of a coordinator. Consider (49a)–(49j):

(49) a. (Either) John will come *or* Carol will go.
b. You like (either) coffee *or* tea.
c. John will (either) smoke *or* drink coffee.
d. (Either) Come in *or* go out! [cf., '(*Either) Come in! *Or* go out!']
e. Do you like (either) coffee *or* tea? [polar, cf., (49g)]
f. (*Either) Homo sapiens *or* modern humans are the last surviving species of the genus Homo.
g. Do you like (*either) coffee *or* tea? [choice, cf., (49e)]
h. (*Either) Do you like coffee, *or* do you like tea? [choice]
i. (*Either) Run, *or* you will be late.
j. (*Either) Mommy must hang up, okay! *Or* she will never come home.[82]

In (49a)–(49e), 'or' is used as a *disjunction marker*, a *coordinator* that conjoins two or more phrases or clauses of the same syntactic category to yield a coordinated phrase or clause (viz., a *disjunction*) belonging to the same category. In

[82] In (49a)–(49j), I make the parenthetical addition '(either)' or '(*either)' to indicate whether or not the plain 'or' in the examples is interchangeable with 'either ... or ...'. But the examples *proper* do not include the parenthetical additions. For example, (49e) is 'Do you like coffee *or* tea?' (without '(either)'), which is homonymous with (49g) (see below for more on this). Incidentally, the last example, (49j), is drawn from the movie *Erin Brockovich*, where Ms. Brockovich (Julia Roberts) says it to her daughter on a pay phone. (I used to use the example in elementary logic classes to illustrate the logical connection between 'or' and 'if' (and 'if ... not ...' and 'unless').)

these cases, the plain 'or' is interchangeable with the emphatic 'either ... or ...', which yields, e.g., 'either coffee or tea'. In (49f), 'or' figures as a coordinator for *apposition*, a marker of *coordinative apposition*, but not as a disjunction marker. This contrasts the plain 'or' with 'either ... or ...', which cannot be used for apposition. How about the use of 'or' in the choice interrogatives (49g)–(49h)?

The choice interrogative (49g) is homonymous with its polar cousin: (49e). But they have different translations in some languages (e.g., Japanese, Korean), and we can consider them as the ambiguous interrogative (26) on different readings: the *polar* and *choice readings*. (49e) (i.e., (26) on the polar reading) is used only for a polar question, and (49g) (i.e., (26) on the choice reading) only for a choice question. So they are tantamount to two different unambiguous interrogatives:

(50) a. Do you like *either* coffee *or* tea? [polar]
b. Do you like coffee, *or* do you like tea? [=(49h), choice]

(49e) is interchangeable with (50a), which is an unambiguous polar interrogative; (49g) with (50b), which is an unambiguous choice interrogative.

To explain the difference between (49e) and (49g), or the ambiguity of (26), it is usual to take 'or' to have two different meanings or uses. Haspelmath, as noted above (§2.1), distinguishes disjunction into two kinds: "standard disjunction" and "interrogative disjunction" (2007, 3f). And he takes 'or' to be used for *standard* disjunction in (49a)–(49e), but for *interrogative* disjunction in (49g) (2007, 25f). On this distinction, the 'or' in choice interrogatives is also a *coordinator*, but differs from the 'or' for the usual, "standard" disjunction in being a marker of a special kind of disjunction. I agree in distinguishing the two uses of 'or'. But I do not think it is used as a coordinator in choice interrogatives.

Like Japanese and Korean choice interrogatives, I think English interrogatives involve *mere juxtaposition* of polar interrogatives. For example, (49h) results from juxtaposing two polar interrogatives:

(51) a. Do you like coffee?
b. *Or* do you like tea?

Then (49g) can be taken to result from removing the shared 'do you like' from the second clause of (49h). So (49g) and (49h) can be taken to result basically from contracting a discourse where (51b) takes (51a) as the antecedent:

(52) Do you like coffee? *Or* do you like tea?

Clearly, 'or' is not used in this discourse as a coordinator that conjoins the antecedent interrogative, (51a), with 'Do you like tea?' Rather, it figures as a *part* of the subsequent interrogative, (51b), as an *adverb* added to 'Do you like tea?' In this use, 'or' is synonymous with 'or else', 'else', 'otherwise', 'if not (so)', etc. So it requires implicit *anaphoric* reference to antecedents. In (52), where it takes (51a) as the antecedent, 'or' figures as a short for the subordinate clause '*if* you do *not* like coffee'. The same holds, I think, for the 'or' in the choice interrogatives (49g)–(49h).

Call the 'or' used interchangeably with the adverbial phrase 'if not (so)' and the like, as in (52), the *adverbial* 'or'. We can see that the adverbial 'or' is used frequently and takes a wide range of antecedents. Clearly, (49j) involves the adverbial use of 'or', as a short for '*if* mommy does *not* hang up'. In (49i), too, 'or' figures adverbially, as a short for '*if* you do *not* run' while taking 'Run!' as the antecedent. So the sentence is an *asyndeton*, resulting from mere juxtaposition of the imperative 'Run!' and the declarative '*Or* you will be late.' In choice interrogatives, as in (52), 'or' takes interrogatives as antecedents.

On this analysis, choice interrogatives derive from juxtaposition of *suitable* polar interrogatives (e.g., (51a)–(51b)), as (49i) does from juxtaposition of 'Run!' and '*Or* you will be late.' This does not mean that the choice interrogative (49g) or (49h), for example, is a device for just asking separate polar questions, those one can ask with two polar interrogatives: (51a) and (51b). One might say 'Run' and continue, e.g., '*Or* you will not run' without giving it as a justification for the command. But juxtaposing a declarative lead by 'or' (or 'and') to the imperative yields a sentence (e.g., (49g)) that one can use to give a command while *adding a justification*. The sentence amounts to, e.g., 'Run, for *otherwise* you will be late', which involves the conjunction 'for' that makes explicit how the two clauses are related. But English has a *conventional* construction involving no conjunction that one can use for the same purpose. For another example, consider *tag interrogatives*, such as 'You like coffee, don't you?' This results from juxtaposing 'You like coffee' and 'Don't you?' Still, it is a single interrogative of a special kind, the use of which does not include the *assertion* that the addressee likes coffee. Similarly, I think, English has a *conventional* way to construct choice interrogatives: juxtaposing polar interrogatives with a suitable placement of the adverbial 'or'. The convention requires some of the juxtaposed interrogatives (e.g., the last of them, e.g., (51b)) to include the adverbial 'or'.[83] This gives the choice interrogatives the semblance of syndetic coordination. Still, 'or' is not used in them as a coordinator any more than it is so used in (49i)–(49j).

Call this analysis the *juxtaposition analysis*. It results from applying the analysis of Korean and Japanese choice interrogatives inspired by Chao's analysis of their Chinese cousins to their English cousins. And the analysis almost directly follows from an obvious analysis of the adverbial 'or' in the likes of (49i)–(49j). Using this analysis, we can take English choice interrogatives to involve mere

[83] Note that it is this conventional requirement that distinguishes the structure of English choice interrogatives from those of Korean and Japanese ones. For Korean and Japanese do not require them to include even apparent coordinators (e.g., the Korean *ani-myen* and the Japanese *sore-tomo*).

juxtaposition essentially for the same reason to take their Japanese and Korean cousins to do so.

Usual analyses of English choice interrogatives, by contrast, take the *coordinator approach*, which takes the 'or' in them to be a coordinator (e.g., one for so-called *disjunction* of a special kind). This approach assimilates the 'or' of choice interrogatives to the 'or' of (the usual) disjunction and severs the obvious connection the former has to the adverbial 'or' in (49i)–(49j). But it is hard to see a good reason to take the choice interrogative 'or' to be closer to the coordinator for disjunction than to the adverbial 'or'.

It is also hard to see how those who take the coordinator approach can explain some interesting constraints on choice or polar interrogatives. Consider (53a)–(53e):

(53) a. Do you like *either* coffee *or* tea? [polar]
b. Do you like coffee, *or* do you like tea? [=(49h), choice]
c. Do you like coffee *or* tea? [=(26), polar, choice]
d. Did John *or* Susan invite Mary? [polar]
e. Is it John *or* Susan who invited Mary? [polar, choice]

While (53c) (=(26)) can be used both for a polar question and for a choice question, as noted above, (53a) cannot. The coordinator approach explains this by distinguishing disjunction into two kinds ("standard" and "interrogative disjunction"): 'either ... or ...' is an unambiguous coordinator (marking only "standard distinction"), while the plain 'or' is an ambiguous coordinator, which can be used for either of the two kinds of disjunction (Haspelmath 2007, 25f). But this does not resolve similar problems that do not involve 'either ... or ...'. While the plain 'or' is ambiguous, (53b) and (53d) are not. One cannot use (53b) for a polar question, nor (53d) for a choice question. What gives rise to this resolution of ambiguity? It is hard to see how one can explain it while taking the 'or' in (53b)–(53e) to be a coordinator.

But the juxtaposition analysis yields a straightforward explanation, which applies to all the three cases: (53a), (53b), and (53d). First, (53a) is a polar interrogative involving the disjunctive phrase '*either* coffee *or* tea', for '*either* ... *or* ...' is a genuine coordinator conjoining the two phrases or clauses introduced by 'either' and 'or'. But it cannot be considered a choice interrogative, for one cannot supplement the adverbial 'or' with 'either' to mark the beginning of its antecedents, as we can see with, e.g., (49i): we cannot add 'either' before the antecedent 'Run' of the adverbial 'or' in (49i) (i.e., '**Either* run, *or* you will be late' is ill-formed). Second, we cannot take (53b) to be a polar interrogative involving the 'or' as a *coordinator* for (standard) disjunction, for the coordinator cannot conjoin two interrogatives: 'Do you like coffee?' and 'Do you like tea?' (53b) can be considered a choice interrogative, for the *adverbial* 'or' can take interrogatives (e.g., 'Do you like coffee?') as antecedents. Finally, (53d) cannot be considered a choice interrogative, for the 'or' in the sentence cannot be considered an adverbial 'or' because 'Did John' (which precedes it) does not form an interrogative that the 'or' can take as the antecedent. By contrast, (53e) can be used for a choice question as well, for the adverbial 'or' can take the preceding 'Is it John' as the antecedent. (53c), too, has a choice reading; on this reading, it is essentially an abbreviation of (53b) and has an underlying structure involving two interrogative clauses:

(54) Do you like coffee, *or* [do you like] tea? [choice]

So 'either' cannot be added before 'coffee' in (53c) on the choice reading for the same reason that it cannot be so added in (53b).

2.5 Chinese Choice Interrogatives

It is usual to take Chinese choice interrogatives to involve a special expression (*háishì*) as a coordinator. But Chao (1955), we have seen (§2.1), presents an alternative analysis, the *juxtaposition analysis*, that takes them to involve *asyndetic* coordination by mere juxtaposition of polar interrogatives. We can fruitfully extend his analysis of Chinese choice interrogatives for other languages: Korean (§2.2), Japanese (§2.3), and English (§2.4). With the examinations we have made of choice interrogatives of these languages to extend his analysis, I think we can provide further support for the analysis.

Usual analyses of Chinese choice interrogatives (e.g., Li and Thompson 1981, 653f; Matthews and Yip 1994/2011, §17.2; Haspelmath 2007, 3f; Biezma and Rawlins 2015, 463f; Erlewine 2014) take the coordinator approach. They take *háishì* (or its Cantonese cousins) to be a *mandatory coordinator* for a special kind of disjunction (e.g., the co-called interrogative disjunction) that amounts to the 'or' of English choice interrogatives. Chao, by contrast, does not consider it a coordinator and takes it to be *optionally* added in choice interrogatives. This raises two issues:

Q1. Is *háishì* a coordinator?
Q2. Is *háishì* mandatory or essential in Chinese choice interrogatives?

These are independent issues. Clearly, 'or' is mandatory in English choice interrogatives. But this does not mean that it is a coordinator, we have seen (§2.4), for it can be used adverbially, as in (49i): 'Run, *or* you will be late.' Conversely, some languages might *optionally* involve a *coordinator* conjoining parts of choice interrogatives. How about the Chinese *háishì*? I agree with Chao to answer both questions in the negative.

Is *háishì* a Coordinator?[84] It is usual to give 'or' as the English translation of *háishì* in choice interrogatives (Kleeman and Yu 2010) and, conversely,

[84] For the Chinese examples given below and discussions on them, I am especially indebted Joe Hung. Despite his patient instructions on delicate details of Chinese, I am solely responsible for any error in the discussions.

háishì as the Chinese translation of 'or' in choice interrogatives (Kleeman and Yu 2010; Chao 1948, 58f; Chao 1968, 265). But this does not mean that *háishì* is a coordinator, for the English 'or' can be used adverbially as well and can be taken to be so used in choice interrogatives (see §2.4). Moreover, *háishì* also has non-coordinator uses. Consider (55):

(55) *Nǐ chī fàn? Háishì nǐ chī miàn?*
you eat rice? <*háishì*> you eat noodles
'Do/Will you eat rice? Or (else) do/will you eat noodles?' [cf., (31)]

Clearly, *háishì* is not used as a coordinator in this discourse. And it can be translated by 'or', as suggested in the translation. The 'or' in the translation is the adverbial 'or', which is interchangeable with 'or else', 'otherwise', 'if not (so)', etc. (§2.4), and the *háishì* in (55) can be replaced with *ruò fēi* 'if not (so)' and (*yào*) *bu rán* 'if not so'. Now, we can juxtapose the two interrogatives in (55) to get (56):

(56) *Nǐ chī fàn, háishì nǐ chī miàn?*
you eat rice, <*háishì*> you eat noodles
'Do/Will you eat rice, or (else) do/will you eat noodles?' [choice. cf., (31).]

And by removing the *nǐ* 'you' in the second clause of (56), we can get Chao's example (1955, 35f):

(57) *Nǐ chī fàn, háishì chī miàn?*[85]
you eat rice, <*háishì*> eat noodles
'Do/Will you eat rice, or (else) [do/will you] eat noodles?' [choice. cf., (31)]

So I think that in (56)–(57), for example, *háishì* does not figure as a *coordinator* conjoining two phrases or clauses, but as an *adverbial* part of their second clauses, which stem from the second interrogative of the discourse (55).

This yields a juxtaposition analysis of Chinese choice interrogatives. For example, (56) and (57) are *asyndetons* resulting from juxtaposing two related polar interrogatives:

(58) a. *Nǐ chī fàn?*
you eat rice?
'Do/Will you eat rice?' [cf., (31)]

[85] Except for the comma, this is the same as Chao's example (31) discussed in §2.1. Essentially the same example (with comma) is given in *Oxford Chinese Dictionary* (2010, Chinese-English Dictionary 282).

b. *Háishì* *nǐ* *chī* *miàn?*
<*háishì*> you eat noodles
'Or (else) do/will you eat noodles?' [cf., (31)]

So choice interrogatives with *háishì* result from juxtaposing one or more antecedent interrogatives (e.g., (58a)) with subsequent interrogatives that include it (e.g., (58b)).[86]

Is *háishì* Mandatory. We can now consider the second issue:

Q2. Is *háishí* mandatory or essential in Chinese choice interrogatives?

Chao (1955) gives a negative answer, as we have seen (§2.1), for one can remove *háishí* from, e.g., (57) and still get a choice interrogative:

(59) *Nǐ* *chī* *fàn,* *chī* *miàn?*
you eat rice, eat noodles
'Do/Will you eat rice, or eat noodles?' [choice. cf., (32)]

And we can consider (59) an asyndeton resulting from juxtaposing two polar interrogatives:

(60) a. *Nǐ* *chī* *fàn?*
you eat rice?
'Do/Will you eat rice?' [=(58a)]

[86] The above formulation of the juxtaposition analysis of choice interrogatives with *háishì* assumes the usual view that identifies it with the English 'or', although it takes *háishì* to amount to the adverbial 'or (else)' rather than the 'or' for disjunction. But I think the Chinese *háishì* is closer to the Japanese *sore-tomo* 'in spite of, instead/rather' and the Korean *chalali* 'rather' than the English 'or (else)' or the Korean *ani-myen* 'if not (so)'. And this relates the use of *háishì* in choice interrogatives to its other uses, which amount to (a) 'still' or 'nevertheless', and (b) 'had better' or 'should rather'. I think the notion of '*rather* A (*than* B)' underlies the three main uses of *háishì*. I leave it for another occasion to elaborate on this view. (Note that this view, too, supports the juxtaposition analysis, for the view also takes *háishì* to be an adverb, not a coordinator).

b. *Nǐ chī miàn?*
you eat noodles
'Do/will you eat noodles?' [cf., (58b)]

Some might object that one cannot get a Chinese choice interrogative with mere juxtaposition of (60a) and (60b), just as one cannot get an English choice interrogative with mere juxtaposition of their English counterparts: 'Do/will you eat rice?' and 'Do/will you eat noodles?' On this view, (59) must be taken to result from removing *háishì* from (57), and this means that *háishì* is implicit in (59) because it is essentially involved in its construction.

Note that this defense of the affirmative answer to Q2 does not undermine the juxtaposition analysis of Chinese choice interrogatives. Some might hold the analysis and yet take Chinese to be like English in having constraints on polar interrogatives that can yield a choice interrogative via juxtaposition: as English requires some of them (e.g., the last one) to include the adverbial 'or', so does Chinese require *háishì* to figure in some of them (e.g., the last one, e.g., (58b) in (56)). But some languages (e.g., Korean, Japanese), we have seen, have no such requirement, and Chinese differs from English in having choice interrogatives that do not include *háishì* (e.g., (59)). Moreover, Chinese has a kind of choice interrogatives that rarely include *háishì*: *A-not-A* interrogatives (e.g., the Chinese translation of 'Do/Will you eat rice, or not?') In these interrogatives, Chao says, *háishì* is "almost always omitted" (1948, 59).[87] So I think Chinese differs from English in *not requiring* polar interrogatives to include some special elements (e.g., 'or', *háishì*) to yield a choice interrogative via juxtaposition.

Now, some might object that Choice interrogatives without *háishì* are exceptions, not the norm. Consider a cousin of (59):

(61) *Nǐ chī fàn miàn?*
you eat rice noodles
'*meant*: Do/Will you eat rice, or noodles?'[88] [× choice. cf., (32)]

(61), which does not have *chī* 'eat' before *miàn* 'noodles', is closer to the English 'Do/will you eat rice or noodles?' than Chao's example (32) (cf., (59)), but one cannot use (61) for a choice question. This poses a challenge to articulating how to construct acceptable Chinese choice interrogatives. But I do not think it shows that the juxtaposition analysis is incorrect. It might perhaps help to show that the use of (32) as a choice interrogative is an exception, but this does not mean that Chinese never allows *háishì*-free polar interrogatives to yield a choice interrogative via mere juxtaposition. For it is normal for *A-not-A* interrogatives

[87] See also Li and Thompson (1981, 535–545), whose examples of *A-not-A* interrogatives include none with *háishì*.

[88] I wish to thank one of the referees for the Proceedings of TLLM 2024 to point out this example to pose a challenge to the juxtaposition analysis.

to be devoid of *háishì*. Neither do I think the challenge that (61) poses is *specific* to the juxtaposition analysis. Consider the result of adding *háishì* to (61):

(62) *Nǐ chī fàn háishì miàn?*
you eat rice <*háishì*> noodles
'Do/Will you eat rice, or [do/will you eat] noodles?' [choice. cf., (32)]

This is a choice interrogative, but it seems that one cannot remove *háishì* to get a choice interrogative. If so, those who take *háishì*-free choice interrogatives to result from their cousins with *háishì* would have to explain the constraints blocking the move from (62) to (61). The challenge they face in undertaking this task, I think, would be comparable to the challenge that defenders of the juxtaposition analysis face to explain why one can remove the subject *nǐ* 'you' in (60b) to get Chao's (32) in juxtaposing it to (60a) but cannot additionally remove the verb *chī* 'eat' to get (61).

Now, it would require close study of a wide range of *háishì*-free choice interrogatives to explain what gives rise to the difference in the uses of (59) (or (32)) and (61). But it might be useful to make a couple of observations germane to the issue.[89] Consider some cousins of (61):

(63) a. *Nǐ chī fàn, miàn?*
you eat rice, noodles
'Do/Will you eat rice, or noodles?' [choice?]

b. *Nǐ chī báifàn, chǎomiàn?*
you eat white.rice, fried.noodles
'Do/Will you eat white rice, or fried noodles?' [choice?]

c. *Nǐ chī fàn ne, miàn ne?*
you eat rice INT, noodles INT
'Do/Will you eat rice, or [do/will you eat] noodles?'[90] [choice]

With suitable intonation, it seems, (63a)–(63c) can be used as choice interrogatives.[91] If so, the reason that it is hard to take (61) as a choice interrogative might be that *fàn* 'rice' and *miàn* 'noodles' would usually be taken to form a

[89] The discussion in this paragraph is quite speculative and rather suggestive. It would require more study to give an adequate account of (61) and of (63a)–(63c) below.

[90] *Ne* is one of the interrogative markers. See, e.g., Li and Thompson (1981, 305–307), who say that it has a complementary distribution with *ma*, another interrogative marker.

[91] Joe Hung thinks that (63c) can definitely be used as a choice interrogative (private exchange). But he has a reservation about the use of (63a)–(63b) as choice interrogatives—hence, the question mark after 'choice' for the two. With the requisite intonation and pause, he thinks that (63a) might be actually used to ask *two* questions, albeit with *implication* of a choice question (and the same holds for (63b)). On the juxtaposition analysis, however, the two questions might be taken to have *merged* into a single choice question. In any case, it suffices for the present purpose

single phrase (conjunctive or disjunctive) in the sentence. For in (63c), for example, the interrogative marker *ne* placed between *fàn* and *miàn* separates them to help the former to be taken to be added to *Nǐ chī* ... (*ne*) 'You eat ... (?)' to form the interrogative clause *Nǐ chī fàn* (*ne*) 'Do/will you eat rice (?)', which yields the choice interrogative reading. (63c), I think, also helps to show that *háishì* is not essential in the construction of choice interrogatives.

References

Aristotle: Aristotle's Categories and De Interpretatione, Translated with Notes and Glossary. Translation by J. L. Ackrill. Clarendon Press, Oxford (1963)

Baek, B.: Oykukeloseuy Hankuke Munpep Sacen [in Korean: Dictionary of Grammar of Korean as a Foreign Language]. Yonsei University Press, Seoul (1999)

Beck, S., Kim, S.-S.: Intervention effects in alternative questions. J. Comparative German Linguist. **9**, 165–208 (2006). https://doi.org/10.1007/s10828-006-9005-2

Biezma, M., Rawlins, K.: Alternative questions. Lang. Linguist. Compass **9**, 450–468 (2015)

Chao, Y.R.: Mandarin Primer: An Intensive Course in Spoken Chinese. Harvard University Press, Cambridge (1948)

Chao, Y.R.: Notes on Chinese grammar and logic. Philos. East West **5**, 31–41 (1955)

Chao, Y.R.: A Grammar of Spoken Chinese. University of California Press, Berkeley (1968)

Dryer, M.S., Haspelmath, M. (eds.): The World Atlas of Language Structures Online (WALS Online), v2020.3. Max Planck Institute for Evolutionary Anthropology, Leipzig (2013). http://wals.info

Dryer, M.S.: Position of polar question particles. In: Dryer and Haspelmath (2013), Chapter 92 (2013). http://wals.info/chapter/92. Accessed 04 Aug 2024

Dryer, M.S.: Negative morphemes. In: Dryer and Haspelmath (2013), Chapter 112 (2013). http://wals.info/chapter/112. Accessed 05 Aug 2024

Erlewine, M.Y.: Alternative questions through focus alternatives in Mandarin Chinese. In: Proceedings of the 48th Meeting of the Chicago Linguistic Society (CLS 48), pp. 221–234. Chicago Linguistic Society, Chicago (2014). https://mitcho.com/research/haishi-cls.html. Accessed 03 Aug 2024

Greenbaum, S.: The Oxford English Grammar. Oxford University Press, Oxford (1996)

Haspelmath, M.: Coordination. In: Shopen, T. (ed.) Language Typology and Syntactic Description, 2nd edn, vol. 2, pp. 1–51. Cambridge University Press, Cambridge (2007)

Hinds, J.: Japanese. Routledge, London (2003)

Horn, L.R.: A Natural History of Negation. University of Chicago Press, Chicago (1989)

Imani, I.: Negation. In: Jacobsen, W.M., Takubo, Y. (eds.) Handbook of Japanese Semantics and Pragmatics, vol. 5, pp. 495–536. De Gruyter Mouton, Berlin (2020)

The Japan Foundation: Basic Japanese-English Dictionary, 2nd edn. Oxford University Press, Oxford (2004)

Jespersen, O.: The Philosophy of Grammar. New York, Henry Holt (1924). Reprinting, The University of Chicago Press, Chicago (1992)

that (63c), which does not have *háishì*, can clearly be used as a choice interrogative. (Hung takes this to be the usual reading while not being sure whether it has a polar reading as well).

Jespersen, O.: Essentials of English Grammar. George Allen & Unwin Ltd., London (1933). Reprint, University of Alabama Press, Tuscaloosa (1964)
Kalish, D., Montague, R., Mar, G.: Logic: Techniques of Formal Reasoning, 2nd edn. Oxford University Press, Oxford (1980)
Kim, S.-S.: Questions, focus, and intervention effects. In: Kuno, S., et al. (eds.) Harvard Studies in Korean Linguistics, vol. 11, pp. 520–533. Harvard University Press, Cambridge (2006)
Kleeman, J., Yu, H. (eds.): The Oxford Chinese Dictionary. Oxford University Press, Oxford (2010)
Lapolla, R.: Chao, Y. R. [Zhào Yuánrèn]. In: Sybesma, R., et al. (eds.) Encyclopedia of Chinese Language and Linguistics, vol. 1, pp. 352–356. Brill, Leiden (2017)
Lee, C.: Contrastive topic and/or contrastive focus. In: McClure, W. (ed.) Japanese/Korean Linguistics, vol. 12, pp. 352–364 (2003). https://semanticsarchive.net/Archive/mQ5MWMyY/jk12Lee.pdf
Lee, I., Chae, W.: Kuke Munpeplon Kanguy [in Korean: Lectures on Korean Grammar]. Hakyensa, Seoul (2000)
Lee, I., Ramsey, S.R.: The Korean Language. State University of New York Press, Albany (2000)
Lewis, C.I., Langford, C.H.: Symbolic Logic. The Century Co., New York and London (1932)
Li, C.N., Thompson, S.A.: Mandarin Chinese: A Functional Reference Grammar. University of California Press, Berkeley (1981)
Martin, S.E.: Essential Japanese, 3rd rev. ed. Tuttle, Tokyo (1962)
Martin, S.E.: A Reference Grammar of Korean. Tuttle, Tokyo (1992)
Martin, S.E.: A Reference Grammar of Japanese. University of Hawaii Press, Honolulu (2004)
Martin, S.E., Lee, Y.H., Chang, S.-U.: A Korean-English Dictionary. Yale University Press, New Haven (1967)
Masuoka, T., Takubo, Y.: Kiso Nihongo Bunpō [in Japanese: Basic Grammar of Japanese]. Kuroshio Shuppan, Tokyo (1989)
Masuoka, T., Takubo, Y.: Tayhak Kicho Ilpone Munpep [in Korean: Basic University Grammar of Japanese]. Korean translation of Masuoka and Takubo (1989) by Lee, H. Sisa Ilboneosa, Seoul (1992)
Matthews, S., Yip, V.: Cantonese: A Comprehensive Grammar. Routledge, London (1994). 2nd edn, Routledge, London (2011)
McCawley, J.D.: Everything that Linguists have Always Wanted to Know about Logic* *but were ashamed to ask. University of Chicago Press, Chicago (1981)
Quine, W.V.: Methods of Logic, 4th edn. Harvard University Press, Cambridge (1982)
Saltarelli, M.: Basque. Croom Helm, London (1988)
Sohn, H.-M.: The Korean Language. Cambridge University Press, Cambridge (1999)
Song, S.C.: Explorations in Korean Syntax and Semantics. Institute of East Asian Studies, University of California, Berkeley (1988)
Strawson, P.F.: Introduction to Logical Theory. Methuen, London (1952). Routledge Revivals, Routledge, London (2011)
TfCS. U.d.: Open and closed-choice questions. Available at a site of Tools for Clear Speech (TfCS): https://tfcs.baruch.cuny.edu/open-and-closed-choice-questions/. Accessed 01 Aug 2024
Uegaki, W.: Japanese alternative questions are disjunctions of polar questions. Proc. SALT **24**, 42–62 (2014)

Uegaki, W.: Cross-linguistic variation in the derivation of alternative questions: Japanese and beyond. In: Crnič, L., Sauerland, U. (eds.) The Art and Craft of Semantics: A Festschrift for Irene Heim, vol. 2 (MIT Working Papers in Linguistics, vol. 71), pp. 251–274 (2014)

Whitehead, A.N., Russell, B.: Principia Mathematica, vol. 1, 2nd edn. Cambridge University Press, Cambridge (1925)

Yan, F.: Muhle Minshue [in Chinese: Mill's Logic]. 2nd edn. Chinese translation of J. S. Mill's System of Logic. Commercial Press, Shanghai (1923)

Author Index

B
Bassi, Itai 93

F
Fang, Jingzhi 23

H
He, Rong 51
Holliday, Wesley H. 1

J
Ju, Chen 127

L
Liu, Mingya 106

M
McHugh, Dean 69

T
Trinh, Tue 93

W
Wang, Wei 127
Wimmer, Alexander 106

Y
Yan, Jialiang 127
Yi, Byeong-uk 149

J. Yan et al. (Eds.): TLLM 2024, LNCS 15410, p. 189, 2025.
https://doi.org/10.1007/978-3-031-86054-6